VISION SAGAR

Indian Air Force in the Indian Ocean

VISION SAGAR

Indian Air Force in the Indian Ocean

Wing Commander Vikas Kalyani

INDIA FOUNDATION
Analysis • Advocacy • Awareness

PENTAGON PRESS LLP

VISION SAGAR: Indian Air Force in the Indian Ocean
Wing Commander Vikas Kalyani

First Published in 2021

ISBN 978-93-90095-27-8

Published by
PENTAGON PRESS LLP
206, Peacock Lane, Shahpur Jat
New Delhi-110049
Phones: 011-64706243, 26491568
Telefax: 011-26490600
email: rajan@pentagonpress.in
website: www.pentagonpress.in

Printed at Aegean Offset Printers, Greater Noida, U.P.

Dedicated to my mother

Contents

VICE CHIEF OF THE AIR STAFF

एयर मार्शल एच एस अरोरा प वि से मे अ वि से मे ए डी सी
सह वायु सेना अध्यक्ष
भारतीय वायु सेना
Air Marshal HS Arora PVSM AVSM ADC
Vice Chief of the Air Staff
Indian Air Force

Tele : (011) 23012579
Fax : (011) 23018981
Resi : (011) 23793048

वायु सेना मुख्यालय
वायु भवन, रफी मार्ग
नई दिल्ली - 1100106
Air Headquarters
Vayu Bhawan, Rafi Marg
New Delhi - 110106

FOREWORD

The Indian Ocean has an important place in Indian civilisations since ages. Thousands of years ago the Indus Valley civilisation, as well as the Indian peninsula, had maritime trade links as well as cultural links with many countries through the Indian Ocean. This ocean has shaped much of India's history and now holds the key to the future of the world. While carrying most of India's trade and energy sources, it is also the lifeline of global commerce and plays an important role in foundation of the 'Asian Century'. Since it connects regions of diverse cultures and bears ships of major powers, it raises concerns of stability, contest and thus security.

Air power has developed from flight of few meters to flights covering the whole globe without landing. Due to its inherent characteristics of reach, speed, responsiveness, mobility, versatility and heavy lift capability, airpower has become preferred military choice for rapid response to any kind of situation both at national and international level. It has become an indispensable tool for furtherance of image of a nation. The strategic reach of the IAF in the IOR has been enhanced with induction of modern platforms, which can react at short notice and deliver suitable deterrence as well as succour. Keeping abreast with the development of air power, its use for pre-emptive action with precision targeting has been aptly displayed by the IAF in the recent past. There also have been instances where India has used its Air Force to safeguard sovereign interest of self or friendly neighbours and has helped in restoring normalcy. The new age IAF has acquired the capability to conduct HADR missions across the globe to promote India's soft power diplomacy. Similarly evacuating Indian diaspora as well as citizens of other countries from conflict zones has brought the IAF much recognition as a capable force. The continued contribution of the IAF in UN peace keeping missions since 1950s has been appreciated worldwide.

SAGAR represents India's policy for regional neighbourhood in the IOR and should be looked at as a source to create new strategic collaborations through invigorated relations. This research by Wing Commander Vikas Kalyani is along a thought process that brings out importance of defence forces, IAF in particular, in contributing to the goal of 'security and mutual growth' of the IOR - the crux of vision SAGAR.

It is time to strengthen India's defence while focusing on increasing role of air power for maritime security in the IOR. Today, India's maritime borders demand greater protection from conventional and non-conventional threats where air power can play vital role towards power projection. This book has brought out the capabilities, experience and potential of the IAF to be an important tool for SAGAR. It has also brought out the need for innovative employment of the IAF in the IOR in bilateral and multilateral defence cooperation as well as engaging other air forces in multi-dimensional security architecture. The need for enhancing jointness in terms of planning, procurement and asset utilisation, which is the need of the hour, has also been suggested by the young scholar. It has also been justified well that the IAF is an inevitable instrument to further military diplomacy in the IOR. This work can be a useful read for all concerned with Indian defence forces, strategic studies, defence policy and planning as well as foreign policy.

(HS Arora)
Air Marshal
Vice Chief of the Air Staff

14 Jan 2021

Preface

In an increasingly inter-dependent world characterised by enhanced economic and trade linkages, the oceans have been and will remain life lines of the world economy. In the 21st century, with the Asian economies faring well even in the troubled times, there is no qualm that the Indian Ocean is at the centre of the emerging 'Asian Age'. The economic prominence of the Indian Ocean and its energetic role in the continued growth and prosperity of the littoral nations is well established. With the growing importance of the Indian Ocean Region (IOR) it can be safely said that Indo-Pacific concept has found a strong base and is gaining significance as the most strategic region of the world.

Cultivating an environment of peace and stability in this region is therefore a priority for all the stake holders. Despite the diversity, the challenges that all nations are facing are fairly alike and hence there should be a vision for the region for cooperation and collective action.

India in the Indian Ocean; be it geo-strategically, geo-politically or geo-economically, has been a key player for many centuries. In recent years, the Indian policies towards the Indian Ocean have undergone major changes bringing out the old values of mutual growth and prosperity. Security And Growth for All in the Region

(SAGAR) is the latest acronym encompassing the responsible attitude of Indian government towards the IOR.

While mutual growth remains the underpinning of all foreign cooperation, this growth cannot be assured without a secured environment. Keeping the word 'Security' of 'SAGAR' in perspective, this research suggests that although Indian Navy is carrying out its assigned role of guarding the seas around India with efficacy, the utmost responsibility of securing vast maritime arena called the Indian Ocean has to include air power prowess of India as well, which primarily lies with the Indian Air Force (IAF). Enhanced jointness in operations, procurement and more importantly, orientation at the national level for fulfilling the aim of a secured IOR and invigorated cooperation at international level to boost overall air power synergy in the IOR has been advocated in this book.

The reason behind this research is curiosity about geo-politics, international relations and defence diplomacy of India which grew during various in-service courses. The exposure to research methodology in Defence Services Staff College (DSSC) has helped in the presentation of my views about India's responsibility and role as a regional security provider in the IOR and the undeniable role of the IAF in achieving this national objective.

At the outset I am thankful to my organisation- the IAF and considerate seniors who understood my grounds and interest both and gave me an opportunity to work with India Foundation – a leading think tank of the country. I show my gratitude to all the members of the Board of Governors and Directors of India Foundation who while giving me all the support and guidance, also gave me huge exposure to the think tank community and a befitting platform for interaction with experts from various fields. Various forums organised by India Foundation like Indian Ocean Conference, India Ideas Conclave, Counter Terrorism Conference, Kautilya Fellows Programme, Young Thinkers Meet, India-ASEAN youth summit, Breakfast Briefings etc. are few examples that enhanced my knowledge immensely. Without this support from India Foundation, I could not have met a large number of eminent personalities from the strategic community and gathered necessary wisdom to carry out this work. Maj Gen Dhruv Katoch SM VSM (Retd) has been a kind guide to me who improved my research skills constantly while giving me complete liberty to put my thoughts on paper. I would also like to thank the colleagues at India Foundation who always helped me in this endeavour.

I am extremely grateful to the Vice Chief of Air Staff, Air Marshal HS Arora PVSM AVSM ADC for being kind enough to pen the foreword of this book. My sincere regards to Air Marshal Sumit Mukerji PVSM SC VSM (Retd) and Air Marshal PK Roy PVSM AVSM VM VSM (Retd) who took out time from their busy schedule to be the discussants of my paper and gave their valuable inputs to refine this paper. This work got value addition with kind inputs by Vice Admiral Shekhar Sinha PVSM AVSM NM & Bar (Retd), Air Vice Marshal SK Jha AVSM and Air Vice Marshal Surat Singh AVSM VM VSM who never hesitated in giving me their precious time. I am also thankful to Mr Amit Narang, Joint Secretary, IOR division of MEA for accommodating an interview in his busy routine.

I take this opportunity to thank my friends who were great positive critics and a greater support in my work. I heartily appreciate kind gesture by Professor Harsh Pant, Angad Singh and Anshuman Mainkar who went through different drafts and gave their views to improve the content. I am equally grateful to Mr Rajan Arya and his team at Pentagon Press LLP to give my work a great shape in form of this book. At the end, my family deserves many thanks for all support despite not getting my full attention during last two years. These acknowledgements will be incomplete without special thanks to my wife, Bhavna for all the love, support and endless supply of coffee with generous doses of criticism. She also took all the pain of going through the final draft of my work and yet survived.

Wing Commander Vikas Kalyani

List of Abbreviations

A2/AD	Anti-Access/Area Denial
AAR	Air to Air Refuelling
AD	Air Defence
ADIZs	Air Defence Identification Zones
ADMM-plus	ASEAN Defence Ministers' Meeting-plus
ALH	Advanced Light Helicopter
AMCA	Advanced Multirole Combat Aircraft
ANC	Andaman and Nicobar Command
ANI	Andaman & Nicobar Islands
AOR	Area Of Responsibility
ARF	ASEAN Regional Forum
ASEAN	Association of Southeast Asian Nations
AWACS	Airborne Warning And Control System
AWDT	Air Warrior Drill Team
BRI	Belt and Road Initiative
C^4ISR	Command, Control, Communications, Computers, Intelligence, Surveillance and Reconnaissance
CasEvac	Casualty Evacuation
CBG	Carrier Battle Group
CBM	Confidence Building Measures
CCATT	Critical Care Air Transport Team
CCP	Chinese Communist Party

CDS	Chief of Defence Staff
CENTCOM	Central Command
CENTO	Central Treaty Organisation
CERT	Computer Emergency Response Team
CGPCS	Contact Group on Piracy off the Coast of Somalia
CPEC	China Pakistan Economic Corridor
CSRS	Coastal Surveillance Radar Stations
DMA	Department of Military Affairs
DRC	Democratic Republic of the Congo
DRDO	Defence Research and Development Organisation
DSA	Defence Space Agency
EAS	East Asia Summit
EEZ	Exclusive Economic Zone
EU	European Union
EW	Electronic Warfare
FRA	Flight Refuelling Aircraft
GPS	Global Positioning System
HADR	Humanitarian Assistance and Disaster Relief
IAC	Indian Aviation Contingent
IACCS	Integrated Air Command and Control System
IAF	Indian Air Force
IDS	Integrated Defence Staff
IFC	Information Fusion Centre
IMAC	Information Management and Analysis Centre
IN	Indian Navy
INDOPACOM	Indo Pacific Command
IOC	Indian Ocean Conference
IONS	Indian Ocean Naval Symposium
IOR	Indian Ocean Region
IORA	Indian Ocean Rim Association
IOW	Indian Ocean World

IRNSS	Indian Regional Navigation Satellite System
JMT	Joint Military Training
KMMTTP	Kaladan Multi-Modal Transit Transport Project
LCA	Light Combat Aircraft
LCH	Light Combat Helicopter
LRMP	Long-Range Maritime Patrol
MAFI	Modernisation of Airfield Infrastructure
MAO	Maritime Air Operations
MARCOs	Marine Commandos
MDA	Maritime Domain Awareness
MEA	Ministry of External Affairs
MMT	Mobile Medical Teams
MNDF	Maldivian National Defence Forces
MoD	Ministry of Defence
MoU	Memorandums of Understanding
MRCA	Multi Role Combat Aircraft
NavIC	Navigation with Indian Constellation
NCE	Non-Combat Evacuation
NDRF	National Disaster Relief Force
OBOR	One Belt One Road
ODAs	Off-shore Development Areas
OOA	Out of Africa
OOAC	Out Of Area Contingency
PACOM	Pacific Command
PASSEX	Passage Exercises
PLAN	People's Liberation Army Navy
ReCAAP	Regional Cooperation Agreement on Combating Piracy and Armed Robbery against Ships in Asia
RPA	Remotely Piloted Aircraft
RSTA	Reconnaissance, Surveillance and Target Acquisition
SAGAR	Security And Growth for All in the Region

SAM	Surface to Air Missiles
SAR	Search and Rescue
SCS	South China Sea
SEAD	Suppression of Enemy Air Defences
SIGINT	Signal Intelligence
SKAT	Suryakiran Aerobatics Team
SLOC	Sea Lane Of Communication
STOL	Short Take Off and Landing
UCAV	Unmanned Combat Air Vehicle
UN	United Nations
UNCLOS	United Nations Convention on the Law of the Sea
UNMIS	United Nations Mission in Sudan
USSR	Union of Soviet Socialist Republics
WMD	Weapon of Mass Destruction
WTO	World Trade Organisation

1

Introduction

The Indian Ocean is the third largest of the world's oceanic divisions covering approximately 73 million sq km (which is 20 per cent of the water on the Earth's surface and two-and-a-half times larger than Africa). It is bounded by Asia on the north, on the west by Africa, on the east by Australia, and on the south by the Southern Ocean or, depending on definition, by Antarctica. First attested in 1515 in Modern Latin (*Oceanus Orientalis Indicus*), named for India, which projects into it, the Indian Ocean was earlier the Eastern Ocean, as opposed to the Western Ocean (Atlantic) before the Pacific was surmised.[1]

India's location at the very centre of the Indian Ocean has linked it with other cultures, shaped its maritime trade routes, and influenced its strategic thought. The Indian peninsula is the biggest landmass jutting into the Indian Ocean, and in that respect, the Indian Ocean is a symbolic acknowledgement of India's role in the Indian Ocean Region (IOR). The waters of the Indian Ocean have not only carried commerce but have borne India's culture, religion, and ideas far and wide. India is home to some of the oldest seaports in the world and historically has had extensive maritime links with

Africa, the Gulf, the Mediterranean, South East Asia and the Far East.[2]

With the rise of Asian economies like India, China and the Association of Southeast Asian Nations (ASEAN) and; the emergence of Africa as the new continent of economic growth, the centre of gravity of global commerce has shifted from the Atlantic to what is now described as the 'Indo-Pacific'. Consequently, the Indian Ocean has again become the crossroads of global commerce.[3] Today, the Indian Ocean is one of the most critical and busiest naval transportation links in the world. Almost a hundred thousand ships pass through it in a year, carrying about half of the world's container shipments, one-third of the world's bulk cargo traffic and two-thirds of the oil shipments. The Indian Ocean has now gained vital importance well beyond the shores of the littoral states.[4] The biggest or at least most accessible oil and gas deposits are located on its periphery along with 40 per cent of all offshore petroleum production. The most populous subcontinent lies on its northern shores while Africa holds some of the greatest reserves of strategic minerals and other natural resources, including peerless, still largely untapped biodiversity which is attracting increasing international attention.[5] While the attention of other countries towards the IOR is fairly recent, India has maintained trade and civilisational links with many countries in the region long before it was a colonial lake.

The economies of many of the littoral countries depend heavily on the commercial ports, the shipping industry, and most importantly, the enormous natural resources that enrich these waters with a profusion of marine life. For India, of course, the Indian Ocean is of vital importance[6] with an extensive coastline of around 7,500 km, including a coastline of 36 islands of Lakshadweep Archipelago in the west and 572 islands of the Andaman and Nicobar Islands (ANI) in the east. India's Exclusive Economic Zone (EEZ) is 2.4 million square kilometres and 90 percent of its trade by volume and almost all of its oil imports come by the sea. While the Indian Ocean is important for India, the contrary is also true. India's role as the key player in the Indian Ocean region for stability, security, peace and mutual growth is imperative, not only geographically but by virtue of a shared historical, trade and cultural heritage that binds India with all nations across these waters.[7] India has always promoted platforms for mutual progress and development of the IOR through politico-diplomatic ties along with ensuring the security of the region by assisting other countries in maintaining peace with the help of its capable defence forces. While the Indian Navy plays the role of flag-bearer in the IOR, the Indian Air Force (IAF) and the Indian Army have done their best as and when required. The IAF has evolved over the years to gain the capabilities of a global air force and is being given role of a preferable medium to project soft as well hard power. Air power projection by the IAF has fulfilled the need of guarding India's land and maritime borders, deterring non-state actors as well as helping friendly neighbours in times of crisis including man-made or natural disasters. The need for an active role of air power in the IOR needs to be discussed in detail because, according to ever-evolving doctrines of defence forces across the globe, air power is and will remain the key to security on the high seas.

Why is Air Power so Important?

IOR—Past and Present: Before talking about the role of India as a whole and importance of the IAF and air power in particular for the IOR, it is important to understand the significance of this region. There are established historical evidences that the Indian Ocean and India have always been inter-dependent. It was a major conduit of trade, commerce, religions and cultures much before the Pacific and Atlantic oceans. It connected the Indian sub-continent with many countries that were dependent on trade winds and established India as the strongest economy of the world in pre-colonial times.

Undoubtedly, the IOR is important to India and India is important to the IOR. After many historical events in which the Atlantic and the Pacific oceans played a significant role, the IOR has regained its importance and is the most strategic economic corridor supporting major economies of the world.

SAGAR—A Vision for Mutual Growth: India has appreciated the importance of the IOR and Prime Minister Mr. Narendra Modi had made it clear in the first year of his first term that India sees the IOR as a hub for development of all its neighbours. The integrated approach towards the Indian Ocean, called **SAGAR** (Hindi for 'sea') which stands for **S**ecurity **A**nd **G**rowth for **A**ll in the **R**egion, was underlined by the Prime Minister in March 2015 at Mauritius.[8] This vision for the region is to safeguard its organic unity while advancing cooperation.[9] The IOR today faces numerous challenges such as maritime terrorism, piracy, EEZ encroachment and drug trafficking. As a major regional stakeholder, New Delhi is attempting to play a larger role in the IOR to address these challenges[10] because the globalisation of the Indian economy has only added to its dependence on maritime trade and as in historical times it wants a peaceful growth of the region. India can use its capabilities for the collective well-being and mutual benefit of its maritime neighbours in building their capabilities. SAGAR is an unambiguous, high-level expression of India's ideas for the IOR. It has inter-connected distinct fundamentals that highlight India's commitment in the Indian Ocean.[11]

Indian Defence Forces and SAGAR: SAGAR has two important key words for the IOR: **Security** and **Growth**. If followed in letter and spirit, this concept can become a very good base for the Ministry of Defence (MoD) in coordination with the Ministry of External Affairs (MEA) to make India play a pivotal role in the IOR which it rightfully deserves. **Security is fundamental** to the 'SAGAR' vision.

To effectively combat transnational security challenges across the Indian Ocean, including those posed by non-state actors, it is important to develop security architecture[12] based on understanding that strengthens the culture of cooperation and collective action.

Another clearly visible building block of the embryonic Indian Ocean security strategy is the response method to address humanitarian crises and natural disasters. In the last few years, India has been promptly responding to humanitarian crises, whether natural or artificial, within the country or in its neighbourhood. Humanitarian Assistance and Disaster Relief (HADR) capability forms an important part of the outreach efforts of India. The Indian defence forces have been involved in the safe evacuation/rescue of many lives in different countries across the IOR. The IAF and the Indian Navy in particular, have been instrumental in timely execution of relief operations abroad while the Indian Army plays a major role in domestic relief operations.

While focusing on security in the region, the priority for any nation is to protect its own sovereignty from any threat and maintain peace in its area of responsibility. However, if a nation is not capable of averting the threat to its peaceful progress in the region then it is incumbent on the other nations of the IOR to come forward and assist that nation. For these kinds of circumstances, to have mutual reliability and to maintain a certain level of trust, defence co-operation is necessary amongst the nations. Such co-operation, along with other confidence-building measures (CBM), becomes the foundation to make the strategically important concept of 'SAGAR' successful.

Aim

The aim of this book is to bring out the potential of the IAF alongwith its inherent air power characteristics of speed, flexibility and swift response as well as experience, which make it a critical and inexhaustible tool in enhancing the strategic footprint of India in the IOR. While securing own maritime areas of responsibility and Sea Lanes of Communication (SLOCs), this foot print will also ensure an air power umbrella for security of common interests and rules of the seas. Given the rapidly evolving geo-political scenario in the IOR, it is imperative for India to engage maximum IOR countries through defence co-operation on all fronts and ensure their sustenance by regular interaction of military representatives. The IAF can support and complement the existing policies of the MoD and the MEA, to enhance India's image in the IOR and establish the trust factor in the region so that India can be seen as a reliable leader of vision SAGAR.

Tasks by the IAF in the IOR and Beyond: Role of the IAF has been unique because of its ability to respond fast and its inherent characteristics of range, mobility and flexibility. Other than countless rescue missions within the country, many nations have benefitted from the HADR and Non-Combat Evacuation (NCE) mission capabilities of the IAF. There are other examples in history when the IAF undertook combat missions to support another nation and successfully played the role of a reliable friendly nation. From UN operations in Congo starting in 1960 to peace keeping operations in Somalia and Sudan, the IAF has been appreciated by the UN in all missions. In 1988, Operation Cactus (Maldives) brought out that the strategic airlift capability of the IAF was central to the swift and efficient response of the Indian military that rendered the operation brief and bloodless. During the tsunami (2004) relief operations, Operation Rainbow for Sri Lanka and Operation Castor for Maldives were conducted by the IAF. Operation Neer 2014 was conducted to alleviate the water crisis at Male. There are many other examples that highlight the role and importance of air power in the maritime domain.

Capabilities of the IAF in the IOR: In recent times, the IAF has shown its capabilities of reaching anywhere within the IOR at short notice and has also undertaken exercises across the Ocean so as to make itself more familiar with the region. In its pan-India exercise, Gagan Shakti-2018, special attention was given to maritime air operations. The IAF conducted extensive missions flown by its multi-role fighter aircraft Su-30s, Mirage 2000s and long-range bombers Jaguars, in conjunction with Poseidon-8I (P-8I) Long-Range Maritime Patrol (LRMP) aircraft of the Indian Navy. They practised delivery of long-range weapons such as the cruise missile BrahMos and anti-ship Harpoon missiles to refine offensive and defensive tactics against maritime targets. The operation was the largest ever series of air exercises conducted by the IAF.[13] It demonstrated a significant operational range with the help of Air-to-Air Refuelling (AAR) by IL-78 Flight Refuelling Aircraft (FRA). These capabilities, paired with a permanent base in the Andaman and Nicobar Islands, provide evidence that India has improved its capacity for maritime air operations and can enhance maritime surveillance in the IOR significantly. These can be seen as decisive steps towards securing vital SLOCs for India and its friendly neighbours in the IOR, as **air power can play a role as a force multiplier** to accomplish strategic objectives.[14]

Role of the IAF in Vision SAGAR: In military diplomacy, the IAF has been a consistent contributor to peacekeeping forces in Congo, Sudan, South Sudan, Somalia and Sierra Leone. It has done multilateral and bilateral exercises with the air forces of many countries like the US, UK, France, Russia, UAE, Oman, Singapore, Australia and South Africa in the last 10 years but the big change that has taken place is the focus on HADR exercises with friendly countries. In the IOR, India has maximum resources and capability for quick response. It intends to do more joint exercises in the future with other IOR countries.[15]

Acquisitions like Rafale, Multi-Role Combat Aircraft (MRCA) and development of indigenous Advanced Multirole Combat Aircraft (AMCA) with increased number of Tejas, the domestically developed Light Combat Aircraft (LCA) as well as Light Combat Helicopter (LCH) will definitely place the IAF on the pedestal it deserves. The helicopter and transport fleets are also being modernised to meet the latest challenges posed by potential adversaries or constantly occurring natural disasters. With growing concerns about China's ambition to exploit the waters well beyond the South China Sea, India needs to have a strong and deterrent air force along with a formidable navy. While the Indian Navy is on the obvious path of being the guardian of the seas around India, maritime security needs all the components of the defence forces actively involved, specially the IAF. The IAF has started focussing on its maritime capabilities well in time. It needs more exposure to overseas operations to gain experience with regional friendly countries so as to meet its objective as well as the objective of vision SAGAR, i.e., to make India a net security provider in the IOR. All it needs is constant will and support from all the role players.

Needless to say, India is well aware of its crucially important geostrategic location in the IOR. Not only for its own secured present but also for a peaceful and progressive future of the region, India is actively working on developing strategic ties with the littoral countries. However, India is not the only country with a presence in this important maritime domain, as is obvious by the recent appearances of submarines and survey ships of the Chinese People's Liberation Army Navy (PLAN) in the Indian Ocean which was not so regular earlier. A more worrisome fact is heavy investments under debt trap policy as well as joint exercises between Chinese and other navies in the Indian Ocean. For the safety of India's maritime interests and security ambitions in the IOR, it must strengthen its relations in the IOR.

Efforts are being made in this direction at the diplomatic level in the form of visits by dignitaries to important IOR countries, strengthening economic and cultural relations and assisting them in their hour of need. However, there seems to be scope of engaging the IOR nations on defence co-operation more actively and assist in creating a secured air power umbrella with Indian aviation assets in the lead role. This does not mean that Indian soldiers, sailors or airmen need to be on foreign soil regularly as part of their active duty but efforts can be made towards increased engagement of the defence forces of India with other IOR countries with the aim of capacity building and capability enhancement. This approach can boost the synergy amongst the IOR nations, and create bonhomie and confidence and thus can inculcate a feeling of a strong and secure IOR on the lines of the vision SAGAR.

NOTES

1. Indian Ocean, www.etymonline.com/word/indian%20ocean (Accessed 24 December 2018).
2. Sushma Swaraj, SAGAR – India's Vision for the Indian Ocean Region, https://indiafoundation.in/articles-and-papers/sagar-indias-vision-for-the-indian-ocean-region/ (Accessed 11 June 2018).
3. Sanjaya Baru, "India and the Indian Ocean Region: The New Geo-Economics", http://www.indiafoundation.in/india-and-the-indian-ocean-region-the-new-geo-economics-by-sanjaya-baru/ (Accessed 11 June 2018).
4. Swaraj, op. cit.
5. Côme Carpentier, *The Indian Ocean: From Backwater to Strategic Epicentre, From India to Infinity* (New Delhi: Har-Anand Publications, 2012), p. 143.
6. 'Preserve organic unity of Indian Ocean Region, advance cooperation – Swaraj', *The Island*, 03 September 2017, http://www.island.lk/index.php?page_cat=article-details&page=article-details&code_title=171146.
7. Swaraj, op. cit.
8. S. Jaishankar, "Foreword", Yogendra Kumar, (ed.), *Whither Indian Ocean Maritime Order* (Delhi: K W Publisher, 2017), p. 5.
9. The Island, op. cit.
10. Pooja Bhatt, "Achieving Indian Air Dominance in The Indian Ocean Region", https://southasianvoices.org/achieving-indian-air-dominance-in-the-indian-ocean-region (Accessed 17 May 2018).

11. Swaraj, op. cit.
12. The Island, op. cit.
13. Bhatt, op. cit.
14. Ibid.
15. B.S. Dhanoa, in a lecture at Vivekananda International Foundation, April 2018.

2

India and the Indian Ocean: A Venerable Bond

Before taking up the importance of the Indian Ocean for India and vice versa, it is necessary to understand the history of the Indian Ocean and its interaction with humankind. It is important to know how the Indian Ocean carved the future of the region that is the present-day IOR. Following this course, one should be able to understand the uniqueness of the Indian Ocean and the role of various cultures and countries in its historic prosperity before it became a colonial lake and later re-emerged as a major global common.

One can start directly from the appearance of early humans which was about 200,000 years ago in the East African Rift valley. There are two theories of their migration from there. The Out-of-Africa (OOA) theory states that *Homo sapiens* spread from Africa into mainland Eurasia. The more recent Southern Dispersal or Coastal Hypothesis instead advocates that modern humans spread along the coasts of the Arabic Peninsula and southern Asia. This hypothesis is supported by mitochondrial DNA research which reveals a rapid dispersal event during the Late Pleistocene period (around 11,000 years ago). This coastal dispersal, however, began in East Africa approximately 75,000 years ago and occurred intermittently from estuary to estuary along the northern perimeter of the Indian Ocean. It eventually resulted in modern humans migrating from Sunda over Wallacea to Sahul (Southeast Asia to Australia).[1] Since then, waves of migration have resettled people and, clearly, the Indian Ocean littoral had been inhabited long before the first civilisations emerged. About 5000–6000 years ago, six distinct cultural centres had evolved around the Indian Ocean: East Africa, the Middle East, the Indian Subcontinent, South East Asia, the Malay World, and Australia; each interlinked to its neighbours.[2]

Globalisation is a recent trend for the rest of the world but not so for the ancient Indian Ocean World (IOW)—a less-used historical term. Food globalisation began on the Indian Ocean littoral around 4,000 years ago. Five African crops-sorghum, pearl millet, finger millet, cowpea, and hyacinth bean somehow found their way to Gujarat in India during the late Harappan time (2000–1700 BCE). Gujarati merchants evolved into the first explorers of the Indian Ocean as they traded African goods such as ivory, tortoise shell, and even paid labour. Broomcorn millet found its way from Central Asia to Africa, together with chicken and zebu cattle, although the exact timing is disputed. Around 2000 BCE, black pepper and sesame, both native to Asia, appear in Egypt, albeit in small quantities. Around the same time, the black rat and the house mouse emigrates from Asia to Egypt. Banana reached Africa around 3,000 years ago.[3]

The importance and antiquity of the Indian Peninsula in the Indian Ocean, can simply be proven by origin of the great ocean's name. The Indian Ocean is named after India (*Oceanus Orientalis Indicus*) since at least 1515 CE. The Sindhu Mahasagara or the great sea of the Sindhu, as called by the ancient Indians, this ocean has been variously called Hindu Ocean, Indic Ocean, Hind Mahasagar etc., in various languages.[4] India's former External Affairs Minister, Smt. Sushma Swaraj, mentioned at the Indian Ocean Conference at Vietnam in 2018 that it should come as no surprise that in Indian scriptures, the Indian Ocean was known as 'Ratnakara'—the creator of gems. The waters of this great ocean were considered as the source of riches and prosperity. The Indian Ocean is a region where some of the largest and smallest nations of the world have **coexisted in harmony**. This harmony is not only because of economic and cultural commonalities, but also due to ideological and civilisational cohesion.[5]

The Indian Ocean, together with the Mediterranean Sea, has connected people since ancient times, whereas the Atlantic and Pacific oceans have played the roles of North-South oriented barriers. The written history of the Indian Ocean, however, has been Eurocentric and largely dependent on the availability of written sources from the colonial era.[6] The truth is that the IOR was a self-sustained, independent and interconnected world much before Vasco da Gama reached the western coast of India, the event that is mostly the beginning of the history of the IOR from the Western perspective. This is the equivalent of telling European history with little reference to Athens, Venice or the Vikings.[7] It is also important to know that while the Indian Ocean has been the pathway of prosperity from Africa to Australia through centuries, India's geographical location along with its cultural and economic weight made it the pivot of the erstwhile IOW.

India and the Arabian Sea (Western Indian Ocean)

Earlier, the growth of civilisation and distinctive monsoon winds made the Indian Ocean indisputably the first centre of maritime activity. The first naval tradition in this region grew up along the coasts of the Arabian Sea and not around the Aegean, as is believed by European writers. Mr. K.M. Panikkar, in his essay on the Indian

Ocean, mentions categorically that long before seafaring developed in the limited Aegean waters, oceanic navigation had become common with the coastal people of peninsular India. And millenniums before Columbus sailed across the Atlantic and Magellan crossed the Pacific, the Indian Ocean had become an active thoroughfare of commercial and cultural traffic.[8]

There are indisputable evidences of close connection between the civilisation of Nineveh and Babylon (Iraq) and the west coast of India. This was possible only through the navigation of the Arabian Sea. It is also established that there was a flourishing trade between the then Levant (Syria) and the Indian west coast. Different items including gold discovered in Mohenjo-daro (3,000–2,500 BCE) point to their origin in extreme south of India and could have come only by sea. Much of the materials unearthed in the relics of the Indus Valley civilisation came from the coast of the Red Sea and it only adds to the fact that India was connected with the outside world by sea links in those ancient times.[9] There are evidences in the Obelisk of Shalmanesar III (860 BCE) of the presence of Indian elephants in the valley of Euphrates. With Egypt, the relations of the West Coast were very intimate from the earliest times. Panikkar quotes Flinders Petrie, who had discovered the portraits of Indian men and women at Memphis, "These are the first remains of Indians on the Mediterranean. We seem now to have touched the Indian colony at Memphis."[10]

Though the early navigators of the coast were of all races including the Phoenicians, the Chaldeans and the Asiatic Greeks, the Hindus had the preponderant share, as much before the rest, they had begun sailing out into the open sea. The Egyptians or the Greeks never sailed the Arabian Sea till Hippalus made the historic discovery of the monsoon winds in 45 CE.[11] But the navigators of India sailed across and discovered Socotra which was also called *Sukh-Dhara* (land of bliss) long before that time and navigated the

Red Sea as well. It may be corroborated from the statement of Strabo that an Indian sailor who was drifting in the Red Sea was taken to Egypt to show them the route to India. This was in 120 BCE, 165 years before Hippalus's discovery.[12] With this, he had achieved a feat much more remarkable than what Vasco da Gama did 1,500 years later. The journey from Aden to the Malabar Coast reduced to six weeks only and consequently from that time onwards regular voyages took place between the Indian coasts and the Red Sea ports.[13] Because of increased trade between Roman-occupied Egypt and India, the Indian Ocean underwent a revolutionary change. The trade thither became very lucrative. The discovery of a large horde of Roman currency in South India, especially on the West Coast, validates the fact that the Indian, Chinese and Arabic worlds engaged actively with the Romans on the trade front.

India was fully equipped to absorb the trade revolution in those times. There are dozens of Indian ports mentioned in history, most of which can be identified even now. Lothal and Bharuch in Gujarat are such examples. At Bharuch, the ships, on arrival from the Red Sea, were met by government pilot boats and moored in regular basins. Numerous ports from Bharuch to Quilon (now Kollam in Kerala) became significant trade markets. A Tamil classic of the first century describes the port of Muziris (now Kodungallur in Kochi) as being packed with ships. Even a Roman temple's ruins were unearthed in that area.[14]

The main races that used the sea at that time were undoubtedly the Hindus, Asiatic Greeks and the inhabitants of the Arabian coastline as stated by Panikkar. He mentioned that the Hindus were already using a magnetic compass known as *Matsya Yantra* for determining direction in the seas. Further, they had developed great skill in building ocean-going ships of great strength and durability. Tamil scholars had counted approximately 1,800 nautical words in that language. It may, therefore, be said that at the beginning of the

Christian era, the Arabian Sea was not only freely navigated by all the sea-going communities on the Indian coast line but had become a regular highway of cultural and commercial movement.[15]

India and the Bay of Bengal (Eastern Indian Ocean)

The evidence of navigation in the Bay of Bengal before the Christian era is even more substantial. The indubitable presence of flourishing Hindu colonies in Malay, Sumatra, Java and even Annam in the first century CE proves that the Bay of Bengal had also been mastered long before that time.[16] Kautilya's *Arthashastra,* an authoritative work on administration which was written in the fourth century BCE, lays down the functions of the Port Commissioner and Harbour Master. The Board of Shipping was one of the important departments of the Mauryan emperors.[17] As the Mauryan ports were mostly on the coast of the Bay of Bengal, there was significant growth of trade and shipping in that area in the fourth century BCE. In the first century CE also, both inscriptional and numismatic evidences show that the Bay of Bengal was the centre of great maritime activity. The Andhra dynasty (which followed the Mauryas) whose territories also bordered the Bay of Bengal has left some remarkable coins with pictures of ships on them. The growth of large Hindu kingdoms and empires in Champa (Siam, now Thailand), Cambodia, Java, Sumatra and in other areas of Indonesia and constantly renovated Hindu and Buddhist culture for a period of at least 700 years (from the fifth to the thirteenth century CE), establish the existence of close relations between India and South East Asia based on uninterrupted sea traffic.[18]

So it can be said that from at least the fifth century BCE to the sixth century CE this naval supremacy rested with the continental powers in India. First the Mauryas and then the Andhras were the lords of the eastern seas. The sovereignty of the eastern seas then passed to the Pallavas as may be inferred from the great influence which this dynasty exercised on the colonial kingdoms of 'Further India'. The inscriptions in Cambodia and Java are in Pallava characters and the architecture of the period is distinctly inspired by Pallava achievements. In fact, Pallava leadership is writ large in the history of Java, Champa (Thailand) and Cambodia in the seventh, eighth and ninth centuries.[19] In those times oceanic travel was very advanced and Indian seafarers not just crossed the Bay of Bengal at its broadest strait but also sailed far out into the Pacific. Later, the Chalukyas, the Pandyas and the Cholas also maintained powerful navies on the eastern coast while the rulers of Malabar exercised naval power over the seas of the western coast. Figure 2.1 shows the Chola territories and their trade routes reaching into the Pacific Ocean.

Another great historical Indian naval power, the Sri Vijaya

Fig. 2.1: Chola and Srivijaya Territories with Trade Routes

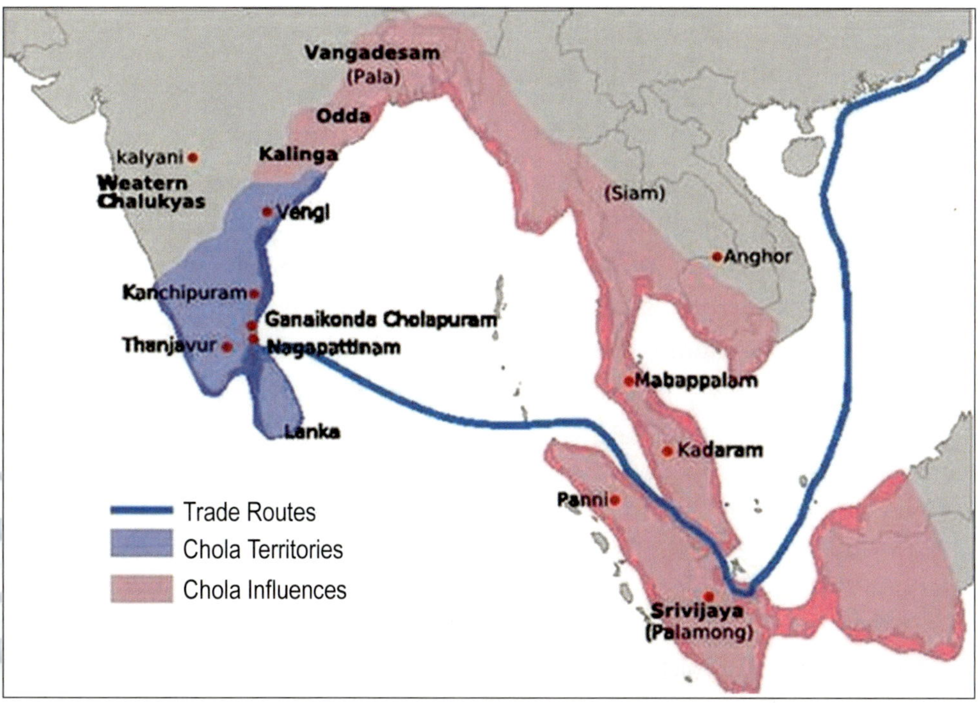

Source: https://pbs.twimg.com/media/DvsdfViUwAA1wBI?format=jpg&name=small

Empire, commanded the Malacca Straits from the fifth to the tenth century. This state included much of peninsular Malaya, Sumatra and the western half of Java. Sri Vijaya kings had a dominating navy which kept the sea clear of pirates and corsairs. With their authority extending over the far-flung group of islands, the Sri Vijaya kings were promoters of Indian seafaring. They were also closely connected with the Indian kingdoms on the western side of the Bay of Bengal, especially with the Kalinga monarchs of Orissa. Until the end of the tenth century, that is for a period of nearly 500 years, the Sri Vijaya kings were the lords of the Ocean. However, in 1007 CE, the Chola Emperor Rajendra fitted out a powerful navy and challenged the might of Sri Vijaya. This naval rivalry lasted for nearly a hundred years. From their profitless overseas adventure, the Cholas finally withdrew by the end of the century. Though it weakened the Sri Vijaya power and opened the way for the Moslem supremacy in Indian waters, the downfall of Hindu naval power did not set in until the destruction of the Sri Vijaya Empire in the 14th century.[20]

After the down fall of the Sri Vijayas and the disappearance of the Cholas from the stage of Indian history, oceanic trade in the Indian seas passed almost exclusively into Arab hands. The Arabs were the great carriers of Indian trade in the fourteenth and fifteenth centuries and their activities extended from the Red Sea ports to Canton (now Guangzhou) and other markets of China. With the Indian monarchs on the coast and with the Indian traders in the ports, they maintained the happiest relations. There was no attempt to exercise naval control at any given time.[21]

The period of Hindu supremacy in the Indian Ocean was one of complete freedom of trade and navigation. While pirates were extirpated and the routes kept open, there was no interference of any kind with trade which was open to ships of all nations. For example, the statement of Abdur Razzak (1442 CE) says that at

Calicut, every ship, whatever place it might have come from or wherever it might be headed, was treated in this port like other vessels and had no trouble of any kind. There was evidently no question of monopoly or exclusion of others from free traffic on the seas.[22] This was the Indian attitude in those times, but with the arrival of the Europeans, the rules of the Indian Ocean started changing.

Indian Ocean Post Vasco da Gama (Short History)

The Portuguese navigator, Vasco da Gama, circled the Cape of Good Hope during his first voyage in 1497 CE and became the first European to sail to India. The Swahili people he encountered along the African east coast had established trade routes to India and China. The Portuguese kidnapped most Swahili pilots in coastal raids and onboard ships. However, few of the pilots were gifts by local Swahili rulers, including a sailor from Gujarat, a gift by a Malindi ruler in Kenya, who helped the Portuguese to reach India. In expeditions after 1500 CE, the Portuguese attacked and colonised cities along the African coast.[23] The reason for this aggression was the Ottoman Empire.

The Ottoman Empire began its expansion into the Indian Ocean in 1517 CE with the conquest of Egypt under Sultan Selim-I. Although the Ottomans shared the same religion as the trading communities in the Indian Ocean, the region was unexplored by them. Hence, they began their own parallel era of discovery which rivalled the European expansion.[24] When the Silk Route (shown in red in Fig. 2.2) was blocked by the Ottoman Empire, the Europeans were in search of a sea route to India and China as the maritime trade in south Asia had been very well established over the course of a few millennia. The route shown in blue was the spice trade route.

Fig. 2.2: The Silk Route and Spice Trade Route

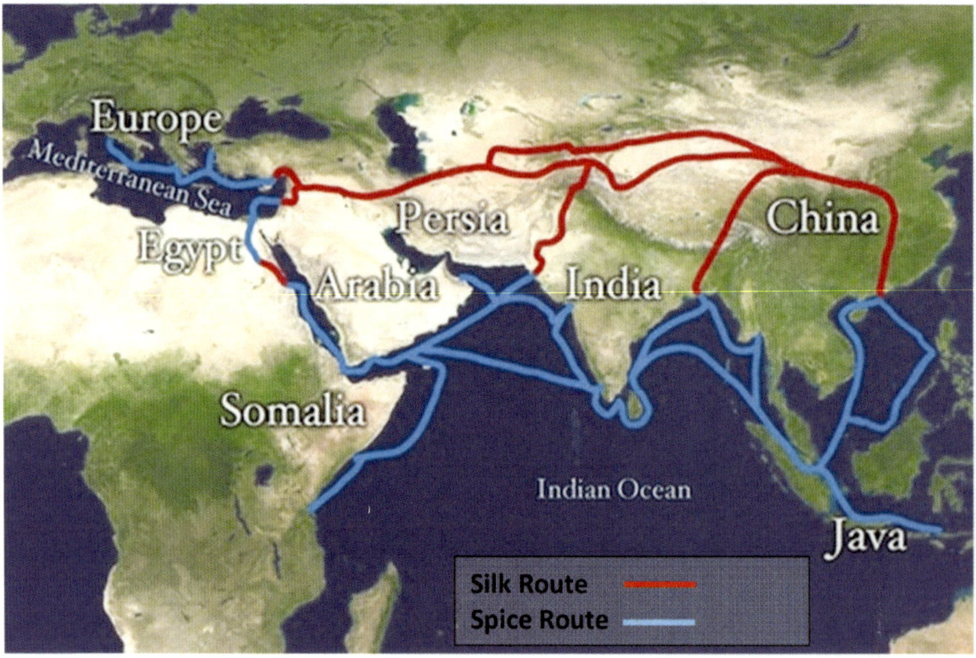

Source: en:Image: Silk Route extant.JPG

The British, Portuguese, French and Spanish started a competition to discover the sea route from Europe to India. Towards end of the medieval period, a vast part of royal finances was spent in building ships to find the sea route to India.[25]

Armed with heavy cannons, the European ships swiftly dominated sea trade. Portugal was the first country which attempted to achieve prominence by setting up forts at important straits and ports. But the small nation was unable to support such a vast project, and they were replaced by other European powers in the mid-seventeenth century. The Dutch East India Company (1602-1798 CE) sought control of trade with the East (mostly Indonesia) across the Indian Ocean. France and Britain established trading companies in the area. Eventually, the Royal Navy became the principal naval power and by 1815 CE, it dominated the area by driving the French out (annihilation of French fleet at Trafalgar in 1805 CE) and making

Indian Ocean a British lake. It was in 1869 that the opening of the Suez Canal made the British grip on the IOR stronger and the annexation of Cyprus made the Mediterranean route to India a private subway for Britain.[26] Two great wars took place in the twentieth century. In the first, Great Britain sailed the Indian Ocean as a supreme power with no nation or group of nations contesting her on the seas. Before the Second World War, Britain developed Singapore as a symbol of its naval might by turning it into a floating fortress. It was a deterrent to anyone who risked questioning the British authority in the Indian Ocean. The rude awakening during the Second World War shook the far-flung branches of the British Kingdom in the IOR and with withdrawal of the Royal Navy, the Indian Ocean for the first time in many centuries, was unguarded in the absence of any dominant maritime power.

The Rude Awakening

"The security of India depended on the mastery of the seas."[27]

Japan introduced strength of **air power** in the maritime domain and proved its importance in control of the seas. The Pacific fleet at Pearl Harbour was destroyed and then the army of Nippon stood on the shores of Johore (Malaysia) overlooking the great fortress of Singapore. In March 1942, Japan overran British Singapore thus threatening the 150-years-old guardianship of the Indian Ocean. Japan's entry into the Indian Ocean clearly challenged India's security. The fall of Singapore was followed by the occupation of the Philippines, Borneo, Sumatra, Java, Malaya and Myanmar by the Japanese forces. Subsequently the Andaman and Nicobar islands were easily captured and the Japanese wanted to prepare them for an aerial assault on India.[28] Ships of the Royal Navy went down at Trincomalee due to an **aerial attack** from carrier-based Japanese planes. Now there was nothing to stop Japan from landing anywhere she pleased on the Indian coast line.[29] However, the timely appearance of the US fleet forced Japan to withdraw from Indian waters. But with this happening, the era of protected sea lanes of communications was over for India and the oceanic areas surrounding India were exposed to despoliation.

Lessons for Air Power Practitioners

It is to the credit of the British that in the early stages of war they learnt the coordination of air and sea operations; and put to devastating effect, the limited air power at their disposal in the battle over the Atlantic Ocean, notwithstanding the disputes between the Royal Air Force and the Royal Navy. It is on record that the RAF played a noteworthy role in maritime operations over the Atlantic. A total of 1,475 enemy surface vessels (representing 1,654,670 tons of shipping) were sunk at sea or destroyed in ports by RAF attacks, constituting 51 per cent of the total enemy losses of 2,885 ships (totalling 4,693,836 tons) destroyed by Allied sea and air action, as also captured, or scuttled from 1939 to 1945. A total of 437 of these ships (186 of which were warships) sank on account of direct air attacks at sea, while 279 others (of which 152 were warships) were bombed and destroyed in ports. Mines laid by Coastal Command and Bomber Command claimed an additional 759 ships, of which 215 were warships.[30]

A post-war study by a joint US Army and US Navy assessment committee of the maritime war against Japan estimates that out of 2,728 Japanese ships sunk in the Pacific Ocean, 1,232 were sunk by direct or indirect air attacks. In concert with other attackers, the air effort sank an additional 46 ships. In all, the air effort accounted for 47 per cent of Japan's maritime losses. In both the Atlantic and Pacific theatres, the key to winning the aircraft-versus-ship war was the achievement of air superiority over the battle area.[31]

After World War II, the Falklands War in 1982 saw offensive use of air power over the sea on a large scale. Both Royal Navy aircraft carriers (some with the Royal Air Force's Harriers on board) and land-based strike aircraft of the Argentinean Air Force (AAF) played important roles in the conflict. While integral air power of the Royal Navy provided air defence to the fleet, its strike capability on board, in terms of the carrier and other ships, was the enabling element in naval and amphibious operations. But what has emerged as perhaps the most notable lesson from the war, was the value of anti-shipping strikes by land-based aircraft. The Argentinean Air Force sank six ships and damaged another 13, thus bringing home the vulnerability of modern ships to such attacks. The damage could perhaps have been greater but for the fact that the AAF was operating at the extreme end of its endurance and did not really have much time available for locating hostile ships. Ultimately, it was the thinnest of margins that separated victory from defeat for the Great Britain.[32]

India and Indian Ocean post World War II

After the end of colonisation in Asia, most of the countries were left with weak economies and power struggles. Also, after World War II, the world witnessed a bipolar system emerging under the Cold War paradigm. Many countries started supporting either of the super powers for survival but India and a number of other countries chose non-alignment as their state policy to avoid the consequences of this model. India also had to be wary of the dangers consequent to the partition, hence security and territorial integrity of the country became paramount.

After the partition, Pakistan claimed parts of Indian territory as its own which resulted in the 1947-48 war. Indian planners saw through these security challenges on the land frontiers as well as in the Indian Ocean, mainly from the two halves of Pakistan in the east and west of India. The Indian Navy's capabilities were being developed for maritime security in two large sections of the Indian Ocean—the Arabian Sea and the Bay of Bengal. Pakistan joined the

US in the Central Treaty Organisation (CENTO) coalition mainly to prevent the Soviet Union from gaining access to warm-water ports in the Arabian Sea. Pakistan started enjoying the US technological support in military hardware. India had the British legacy and then had to look up to the erstwhile Union of Soviet Socialist Republics (USSR) for developing deterrence.[33]

Subsequently, two major wars that India faced were in 1962 against China and in 1965 against Pakistan. Key shortfalls were assessed and defence capability enhancements began sincerely in terms of procurements of fighter aircraft, ships and army weapons. As a consequence, the 1971 war was fought with much domination. The Army thwarted the land attack on the western border efficiently with coordinated operations with the IAF and did not let Pakistan gain any advantage. The IAF achieved total control of the air (air supremacy) in the Eastern theatre as per its planned objectives and this enabled the subsequent blitzkrieg-type operations resulting in the fall of Dhaka on 15 December.[34] It made Pakistan surrender with the capture of more than 90,000 of its troops. The efficacy of the Navy's carrier-borne aircraft was also proven when repeated attacks by the fighter aircraft from sea not only destroyed ships in erstwhile East Pakistani harbours but also blockaded the ports, precluding logistics support to troops and cutting off their escape route.[35] These operations were successful under the greater control of the air generated by the IAF. The importance of the maritime domain around India thus started gaining the attention of policy makers and strategists in India.

Present Day IOR

Today, the IOR is witnessing major shifts in global geopolitics and has found the US as a complete replacement of the United Kingdom. The US correctly assessed the growing importance of the IOR and declared a policy of becoming a pivot to Asia and made India its strategic partner. All this was triggered by the rise of another player in the domain—China. The assertive growth of China in the maritime space had gone unnoticed while the US was engaged in chasing terrorist organisations and individuals. China has utilised vacuum in the Indian Ocean to build its economic and military capabilities. Economically, China has challenged the US for world leadership and militarily, it is threatening the regional countries in the South China Sea (SCS). It has occupied a few structures in the Spratly and Paracel groups of islands and also has made historical claims of vast EEZ nearly in all of the SCS. China claims sovereignty over more than 80 per cent of the SCS, while Vietnam claims sovereignty over the Paracel Islands and the Spratly Islands. The Philippines asserts ownership of the Spratly Archipelago and the Scarborough Shoal, while Brunei and Malaysia have claimed sovereignty over southern parts of the sea and some of the Spratly Islands.[36] China has heavily militarised these sites in order to widen its control on a larger expanse in the SCS. These disputed claims in the SCS can be understood in Fig. 2.3.

Strategic Concerns for India in the IOR

China is now increasing its economic influence on the IOR littorals. The Hambantota port in Sri Lanka was developed by China. Maldives has leased out a number of islands to China for port and airport development and Myanmar also has leased the Kyaukpyu deep water port for development by China. It has also built an 85-metre jetty, naval facilities and major reconnaissance and electronic intelligence systems on the Great Coco Island, which is only 18 km from the northern part of the Andaman Islands of India. However, the presence of these facilities is yet to be confirmed. China is also assisting in constructing a naval base in Sittwe, a sea port of Myanmar, which is close to Kolkata.[37] China has not limited itself to the Bay of Bengal or the Southern Indian Ocean. It has reached the

Fig. 2.3: Disputed Claims in the South China Sea

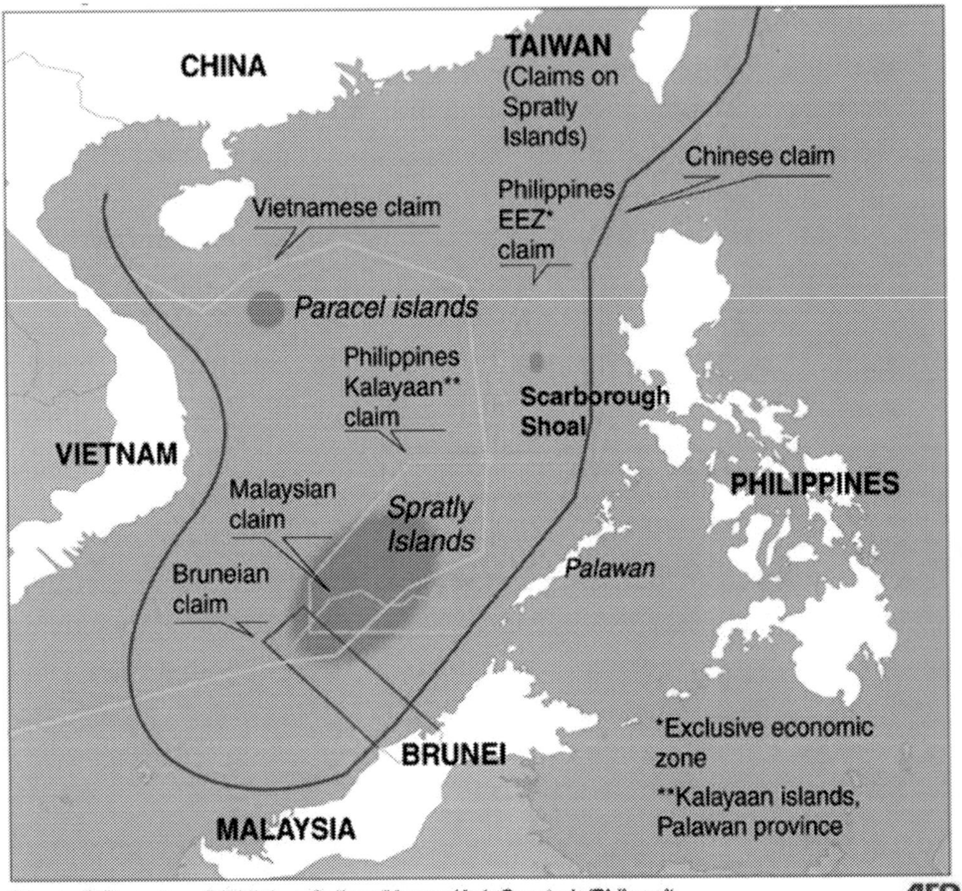

Source: D.Rosenberg/MiddleburyCollege/HarvardAsiaQuarterly/Phil gov't

East African coast as well. China's building of naval base in Djibouti and taking over Gwadar on an operational lease is a clear indication that it has arrived at all the choke points through which not only her trade travels but the entire world's trade is moved.[38] China plans to build a naval and air base on Pakistan's south-western Jiwani Bay, just across the border from Iran and nearly 500 kilometres from the Strait of Hormuz. About 60 kilometres east of Jiwani is Gwadar, where Beijing is developing a massive commercial port at the end of its China-Pakistan Economic Corridor (CPEC). A base at Jiwani would allow China to overlook India's maritime corridor to Afghanistan and Central Asia.[39] China gives this outreach the cover of One Belt One Road (OBOR) or now called the Belt and Road Initiative (BRI). Strategists of India and other concerned countries consider it as a 'string of pearls' strategy with the aim of encircling India and making prominent footprints in the IOR. The CPEC is China's investment scheme in Pakistan which encircles India from the north and is seen clearly as China's contain-India policy. Both 'string of pearls' and CPEC are depicted in Fig. 2.4.

Fig. 2.4: String of Pearls and CPEC

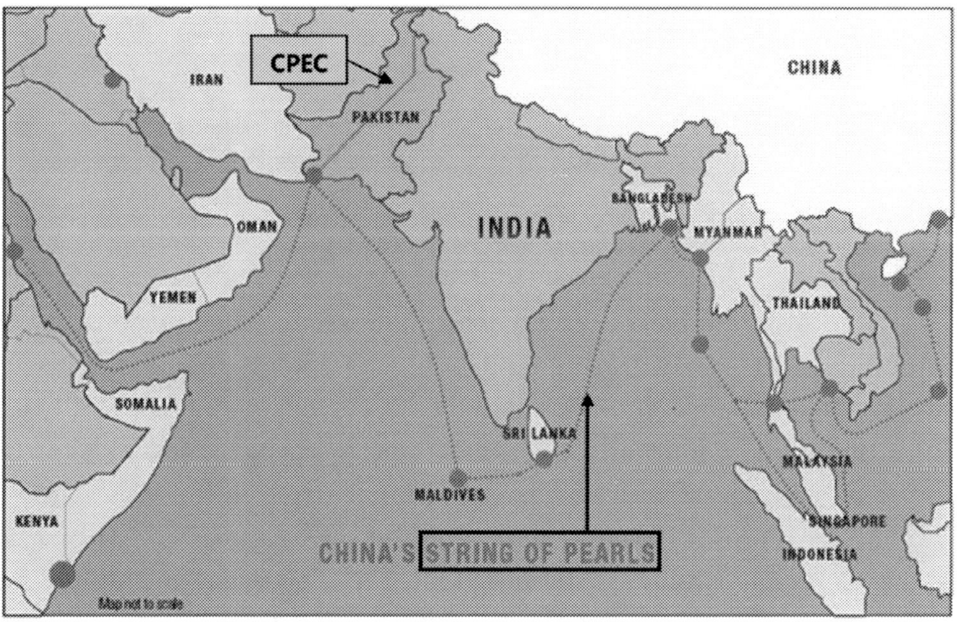

Source: http://www.spsnavalforces.com/story/?id=332

The Chinese map of the OBOR (Fig. 2.5) shows that it is a Chinese version of 'String of Pearls' strategy.

Fig. 2.5: OBOR/BRI: Chinese Map

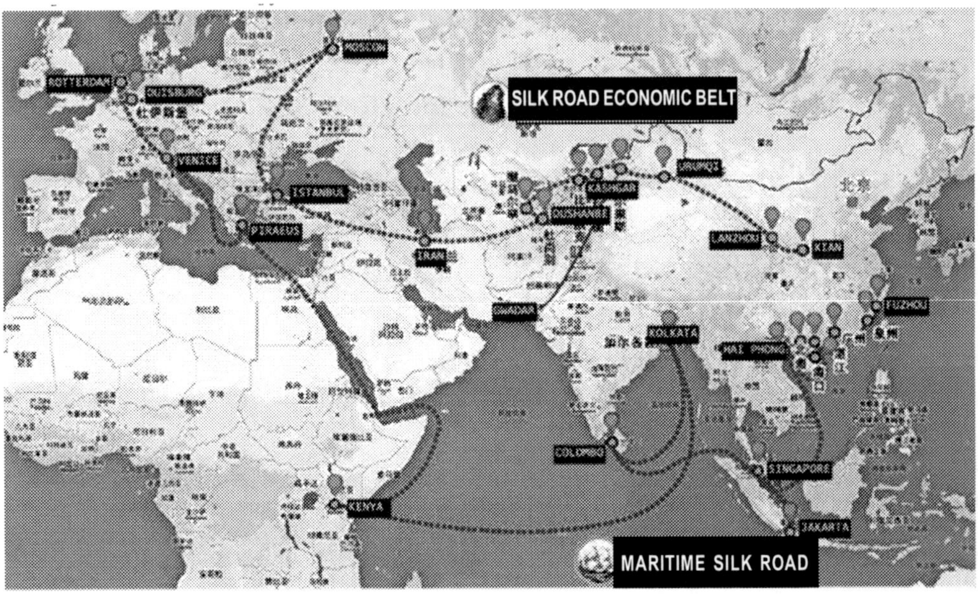

Source: uploaded by Hans-Dieter Evers, https://www.researchgate.net/figure/OBOR-One-Belt-One-Road-Chinese-Map_fig3_319585184

India should be concerned about two issues. One, the security of its maritime trade through the IOR (choke points and SLOCs) as the power balance conundrum in the IOR is picking up (especially in the western IOR) and two; the Chinese assistance to Pakistan which can harass India in its operational space in the IOR. The strategic maritime space of India is shrinking because of these two issues. While China's economic debt trap diplomacy in the IOR littorals is constantly enhancing its influence under its BRI strategy, its arming of Pakistan reflects her desire to keep India engaged away from the Indian Ocean which should be India's main focus area for sustained growth.

Importance of Air Power

In response to China's strategy, America has developed new security architecture in East and South East Asia. It will complement

Washington's designs for an Air-Sea Battle doctrine (later renamed the Joint Concept for Access and Manoeuvre in the global commons) as a counter to China's rising techno-military power in the region. According to this doctrine, the US forces would have to penetrate China's **airspace**, take out Chinese **air defences** and offensive missile capabilities, and then move the US aircraft carrier battle groups closer to China's shores[40] before the aircraft carriers are vulnerable to long-range air/land launched weapons. This shows the importance of air power components in overall maritime defensive or offensive strategies and the dependence of aircraft carriers on air defence capabilities of the nation. But in the Indian context, a joint approach to air warfare has been more oriented to land-air warfare with land-based air forces playing a very limited role over the sea and consequently this role has received less than the desired attention or focus.[41]

Indian planners are cognisant of the emerging geopolitics in the IOR and of the need to strengthen India's posture in the region. Some Indian leaders have drawn a close connection between India's maritime ambitions and its destiny as a great power. Mr. Pranab Mukherjee, former president of India, mentioned that after a very long time of inward and landward focus, India was turning its focus outwards and seawards which is a natural direction for a nation seeking to re-establish itself not only as a continental power but as a maritime power with significance on the world stage.[42] The efforts should be in the direction of India taking the lead role in securing the Indian Ocean's strategic environment. This posture is acknowledged by countries like the US, France and many IOR nations. Many countries are engaging India in bilateral exercises and other defence cooperation. Indian efforts to coalesce the IOR are aimed at keeping the region free of external intervention and conflict.[43] The Indian Navy, which has been considered the flag bearer of India's maritime diplomacy for all these years, is stretching out in the wider Indian Ocean to increase its operational bubble. The Indian Navy ships, submarines and aircraft are deployed for longer durations. But China's military upsurge and imminent basing in the Indian Ocean at places like Djibouti and Gwadar and possible support facilities at ports in Maldives, Sri Lanka and Myanmar, point to the need of strengthening the present IOR policy in terms of using the IAF as well as Army actively for maritime operations and enhanced defence cooperation with maximum IOR countries.

China has started exploring the IOR at an unexpected magnitude and frequency as a result of which some Western thinkers have suggested closer security collaboration with leading stake holders in the region. India's response strategy however must not be based on Chinese actions but on the realisation that growing competition between major powers will put the future of the IOR and nations deriving their peaceful growth from it, in jeopardy. That is where India needs to **think rationally** and **act responsibly** to secure the position of leading nation for the overall growth of the IOR—like the one it had for thousands of years in the past.

NOTES

1. David Bulbeck, "Where River Meets Sea: A Parsimonious Model for Homo Sapiens Colonization of the Indian Ocean Rim and Sahul", *Current Anthropology*, Vol. 48, No. 2, University of Chicago Press, (2007). https://www.jstor.org/stable/10.1086/512988 (Accessed 19 September 2018).
2. McPherson K., "Cultural Exchange in the Indian Ocean Region", *Westerly Magazine*, Vol. 29, No. 4 (1984). https://westerlymag.com.au/cultural-exchange-in-the-indian-ocean-region/ (Accessed 20 September 2019).
3. N. Boivin, Crowther A, Prendergast M, Fuller D Q, "Indian Ocean food globalisation and Africa", *African Archaeological Review*, Vol. 31, No. 4 (2014), pp. 547–581. https://doi.org/10.1007/s10437-014-9173-4.
4. Douglas Harper, "Indian Ocean", *Online Etymology Dictionary*.
5. Sushma Swaraj, https://www.mea.gov.in/Speeches-Statements.htm?dtl/30327/Remarks+by+External+ Affairs+Minister+at+the+3rd+Indian+Ocean+Conference+Vietnam+August+27+2018 (Accessed 20 September 2018).
6. P. Parthasarathi and Riello G., "The Indian Ocean in the Long Eighteenth Century", *Johns Hopkins University Press*, Vol. 48, No. 1 (Fall 2014), pp. 2–3.

7. Sanjeev Sanyal, *The Ocean of Churn* (Gurugram: Penguin Books, 2016), p. 8.
8. K.M. Panikkar, *India and the Indian Ocean: an essay on the influence of sea power on Indian history* (London: George Allen and Unwin Ltd., 1945), p. 23.
9. Ibid.
10. Ibid., p. 24.
11. Ibid.
12. Ibid.
13. Ibid., p. 25.
14. Ibid., p. 26.
15. Ibid.
16. Ibid.
17. Ibid., p. 30.
18. Ibid., p. 31.
19. Ibid., p. 32.
20. Ibid., p. 34.
21. Ibid., p. 36.
22. Ibid., p. 35.
23. Andrea Felber Seligman, "Ambassadors, Explorers, and Allies: A Study of African-European Diplomatic Relationships, 1400-1600", *CUREJ: College Undergraduate Research Electronic Journal*, University of Pennsylvania (23 March 2007). http://repository.upenn.edu/curej/55, pp. 90–95.
24. G. Casale, 'The ottoman "discovery" of the Indian ocean in the sixteenth century', Seascapes: Maritime Histories, Littoral Cultures, and Transoceanic Exchanges, University of Hawai'i Press (01 December 2007), pp. 87-104.
25. Shrey Kapoor, "Why was the Indian Ocean named after India?", https://www.techphlie.com/2016/05/why-was-indian-ocean-named-after-india.html (Accessed 25 July 2019).
26. Panikkar, op. cit., p. 72.
27. Ibid., p. 81.
28. Sanat Kaul, *Andaman and Nicobar Islands-India's Untapped Strategic Assets*, (New Delhi: Pentagon Press, 2015), p. 38.
29. Ibid.
30. H.S. Ahluwalia, "IAF in Maritime Operations", http://bharat-rakshak.com/cms/srr/64-17.htm. (Accessed 05 September 2019).
31. Ibid.
32. Ibid.
33. Shekhar Sinha, "Indian Navy: Opportunities in Changing Geopolitics of IOR", https://www.indiastrategic.in/2018/12/19/indian-navy-opportunities-in-changing-geopolitics-of-ior/. (Accessed 12 September 2019).
34. A.K. Tiwary, "1971 Air War: Battle for Air Supremacy", http://www.indiandefencereview.com/spotlights/1971-air-war-battle-for-air-supremacy/ (Accessed 12 September 2019).
35. Ibid.
36. "Explained: South China Sea dispute", https://www.scmp.com/week-asia/article/2186449/explained-south-china-sea-dispute (Accessed 16 February 2019).

37. Geng Lixin, "Sino-Myanmar Relations: Analysis and Prospects", *The Culture Mandala*, Vol. 7, No. 2 (December 2006). http://www.international-relations.com/CM7-2WB/Sino-Myanmar.htm. (Accessed 20 March 2019).
38. Sinha, op. cit.
39. Farzin Nadimi, "A China-Pakistan Base Deal Could Put Iran on the Back Foot", 27 March 2018, https://www.washingtoninstitute.org/policy-analysis/view/a-china-pakistan-base-deal-could-put-iran-on-the-back-foot (Accessed 25 March 2019).
40. Sinha, "The Indian Navy and the Geopolitics of the Indian Ocean", https://sldinfo.com/2018/12/the-indian-navy-and-the-geopolitics-of-the-indian-ocean/ (Accessed 14 June 2019).
41. Ahluwalia, op. cit.
42. David Brewster, "An Indian Sphere of Influence in the Indian Ocean?", *Security Challenges*, Vol. 6, No. 3 (2010), pp. 1–20. www.jstor.org/stable/26459796 (Accessed 17 June 2019).
43. Arun Prakash, "Introduction", Bhaskar C Uday and Agnihotri Kamlesh K (Eds.), *Security Challenges Along the Indian Ocean Littoral-Indian and US Perspectives* (New Delhi: Matrix Publishers, 2011), p. 3.

3

SAGAR: India's Vision for IOR

A nation cannot survive without a vision. A national vision can be defined as a perspective or framework within which plans (strategies) can be carved out to achieve the national objectives related to growth and security. It lays the foundation of a country's intentions in domestic and international matters, from which flow the policies which become the basis of doctrines of various institutions or organisations within the country. A Doctrine gives birth to strategy and funnels into operations and tactics. But in the process of intentions to become actions, implementation of policies through carefully applied strategies is very important. SAGAR is one such vision which projects the Indian intention of making IOR the most prosperous region in the world. It needs both careful carving of policies and appropriate implementation.

For a long time, India was made oblivious to its maritime legacy and its benign maritime vision. The long colonial period made India a land centric country. Post-independence, the navy was not prepared for handling such big Area Of Responsibility (AOR). The question of sea power did not arise as the Indian Ocean was a British lake for over 150 years. And when the British era ended India was left with troubled land borders to keep all eyes northwards. In 1950, India's share of world merchandise trade was 2.0 per cent, compared to 1.3 per cent for China. By 1990, India's share was down to 0.5 per cent, while China had inched up to 1.8 per cent. By 2010, the relevant numbers were 1.4 and 10.4, respectively.[1] This history which shaped India's vision to be restricted for a long period has been discussed already and it is time to go beyond this period and reacquire India's maritime legacy. This is where SAGAR can play a very important role.

Times have changed and with globalisation, Indian economy is no more inward or insular, but is linked with many of the world's economies. India has embarked on podium of major economies and most of its trade and energy supply depends on SLOCs of the IOR. Once again India from land based economy has become a maritime based economy. The aspiration of becoming a USD 5 trillion economy is on planning charts and undoubtedly the navigation has to be through choppy seas.

The present time is colloquially being called the Asian Century for all the right reasons. If 19th century was Britain's Imperial century and 20th was American, then 21st century is becoming Asian century because of growing influence of Asian countries in the global market. By 2020 the economic growth (in purchasing power parity terms) of Asian giants put together will surpass rest of the world, for the first time since the 19th century[2] and that is going to be future trend as well (as shown in Fig. 3.1).

Hence the interest of extra regional players in IOR is obvious because the future of global commerce lies in the Indian Ocean. Much like the Pacific Rim which shaped the global economic trajectory for two decades, the Indian Ocean has the potential to become the most important source of new global growth over the next 20 years.[3] The IOR has an important role to play for regional peace and

Fig. 3.1: Share of World GDP

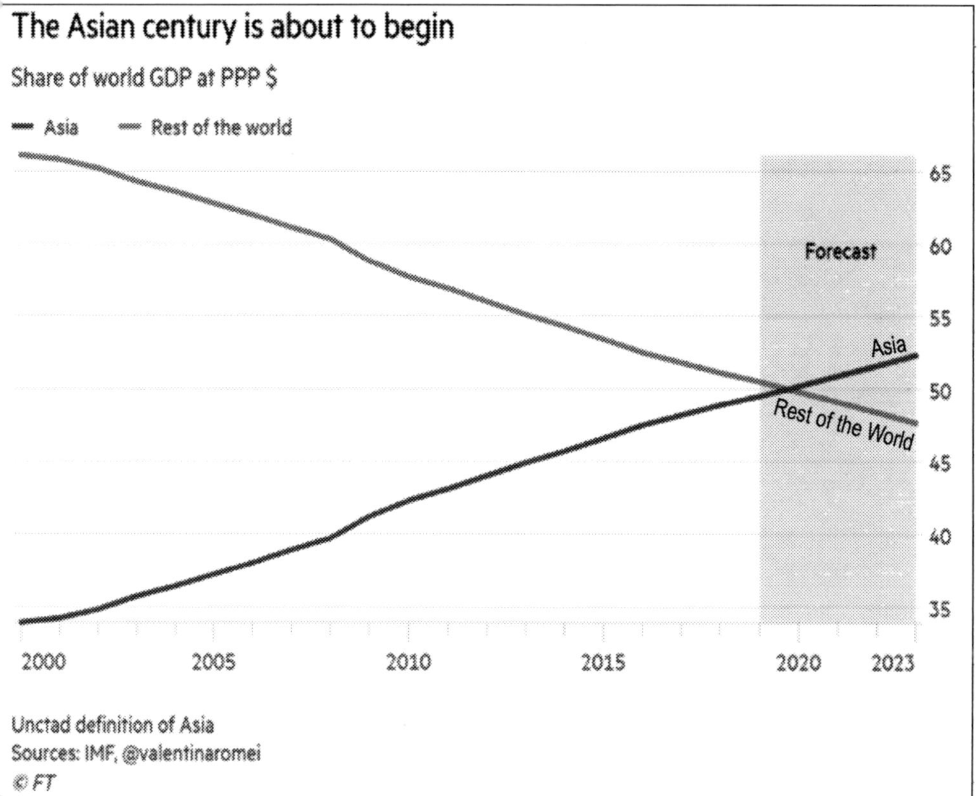

prosperity as the "road of development in the twenty-first century." Today, Indian Ocean has become a major conduit of international trade as it was till few centuries back. While occupying almost 20 per cent of the earth's surface, IOR is inhabited by 35 per cent of world's population staying in 38 littoral states. Over 100,000 ships transit through the Indian Ocean every year, accounting for 66 per cent of the world oil cargo, 50 per cent of container cargo and 33 per cent of bulk cargo.[4] India's ocean borne trade which constitutes 95 per cent trade by volume and 70 per cent trade by value is dependent on SLOCs of Indian Ocean. India's energy security depends on these maritime lifelines. Over 70 per cent of India's oil is imported and this will rise to 90 per cent by 2020.[5] Figure 3.2 brings out importance of the IOR in terms of being major economic corridor.

Fig. 3.2: Major SLOCs Through the IOR

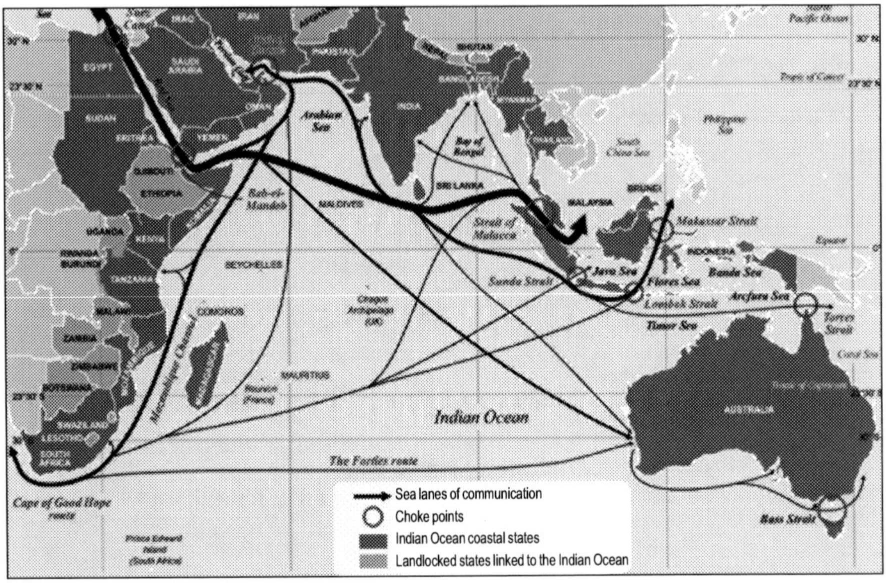

Source: https://books.openedition.org/cei/docannexe/image/469/img-3.jpg

Sitting on the roof of such an important growth hub of the world, India geographically, historically, economically, politically and strategically as well as culturally becomes the central stake holder of the IOR. India has always believed in concept of mutual growth. It follows the principle that when all members of region do well individually, the whole region will grow exponentially and in this process everyone has to pitch in. One nation cannot keep growing alone and ignore the plight of its neighbours. Keeping this outlook as foundation, Indian PM Mr Narendra Modi coined the acronym **SAGAR** for outlining **India's vision for IOR** (complete text of the speech is placed at appendix 1). It is being considered as an articulation of India's national vision about stake in the future of the Indian Ocean. This integrated approach of India has a few key elements as under:

- Do everything to safeguard India's mainland and islands, defend its interests. To ensure a safe, secure and stable Indian Ocean that delivers the shores of prosperity. Make the capabilities available to others who are struck by the natural disasters or any distress.
- Deepen economic and security cooperation with maritime neighbours and strengthen their maritime security capacities and their economic strength.
- Envisage collective action and cooperation to advance peace and security and response to emergencies through multilateral diplomacy and bilateral cooperation. To support the efforts to strengthen India's regional mechanisms for maritime cooperation.
- Seek a more integrated and cooperative future for the region that enhances sustainable development for all. To promote greater collaboration in trade, tourism and investment; infrastructure development; marine science and technology; sustainable fisheries; protection of marine environment; and, overall development of Ocean or Blue Economy.
- Primary responsibility for peace, stability and prosperity in the Indian Ocean rests with the littoral states themselves.[6]

While stating out this vision in Mauritius in March 2015, Mr Modi said, "Indian Ocean Region is at the top of our policy priorities. To me the blue chakra or wheel in India's national flag represents the potential of Blue Revolution or the Ocean Economy. That is how central the Ocean Economy is to us. We often define regional groupings around landmass. The time has come for a strong grouping around the Indian Ocean. We will pursue this with new vigour in the years ahead".[7] The blue economy looks at the oceans as instruments of development and source of livelihood, but cooperation is the key to ensure that the use of oceans is sustainable for long term. This applies to the Indian Ocean in particular as a large number of countries are dependent on its waters. The tsunami of 2004 underscored the need for mechanisms to deal with natural and man-made disasters in the region. The vision SAGAR acknowledges this broader concept of a cooperative maritime order.[8]

Vision SAGAR can prove to be a comprehensive approach for overall development for any Indian Ocean country which adopts it in true sense. For India it is a guiding principle for internal growth as well as a bonding factor with IOR nations for mutual development in a safe and secure environment. This approach needs coordinated implementation within the country as well as in the IOR with other countries. India's former External Affairs Minister, Smt. Sushma Swaraj explained this vision as a process in the Indian Ocean Conference (IOC) organised by India Foundation[9] at Vietnam.

She introduced the three steps involved in the implementation which are: (a) projects to promote hinterland linkages and strengthen regional connectivity, (b) linking South Asia to South East Asia as part of Act East policy and to the Gulf under Think West policy, and (c) playing an active and constructive role in strengthening regional maritime security.

The first part is about India's focus on developing hinterland linkages & regional connectivity. Under the Government's 'Sagarmala' project, initiatives taken include building new ports, modernising old ports, developing inland waterways and hinterland development, which are all aimed at a robust maritime logistics infrastructure. The eastern seaboard is a particular focus and can help recreate an integrated hub and spoke model for regional connectivity in the Bay of Bengal as it was in earlier times. Carrying this focus beyond her borders, India today is devoting more resources and assigning greater priority to building connectivity, contacts and cooperation in her immediate neighbourhood. This is evident in projects in sectors ranging from rail and road transport to power generation and transmission, from port and waterways transport to educational and health exchanges.[10]

The second element as told by Smt. Swaraj was the expanded interpretation of what constitutes India's neighbourhood. This is reflected in the renewed emphasis in government's "Act East" Policy and the new "Think West" policy towards West Asia and Gulf region. The Act East Policy is at the heart of eastward orientation and merges with broader approach to the Indo-Pacific. Over the years, this approach to the region has matured into a broader strategic engagement—with the ASEAN and its related frameworks like the ASEAN Regional Forum (ARF), East Asia Summit (EAS) and ASEAN Defence Ministers' Meeting-plus (ADMM-plus) as also with countries further east, including Japan, South Korea, Australia and the Pacific Islands. India therefore accords high priority to key infrastructure projects such as the Kaladan Multi-Modal Transit Transport Project (KMMTTP) that links Kolkata to Sittwe Port in Myanmar, and the Trilateral Highway that will extend to Thailand through Myanmar. The agreement with Indonesia to develop port infrastructure in Sabong is yet another step in this direction. India is also looking towards a more sustainable future for this region, by collaborating with regional partners on Blue Economy projects, harnessing renewable energy, investing in development of desalination technologies, harvesting the biodiversity of the oceans, and sustainably mining the ocean depths for marine minerals.[11]

She also said, "In all these engagements, we are guided by the development and security priorities of our partners. Our approach is based on inter-dependence rather than dominance or narrow reciprocal considerations. We support responsible and transparent debt financing matched by responsible lending practises. We understand that following universally recognized international norms, transparency, openness, financial responsibility, and promoting a sense of local ownership are essential for better and more sustainable development outcomes".[12]

The third element is contributing to regional maritime security. India is working to ensure the safety and security of maritime traffic through the ocean by strengthening skills and logistics of its Indian Ocean neighbours. It is helping its maritime neighbours set up their coastal surveillance networks for developing shared Maritime Domain Awareness (MDA).[13]

India has signed White Shipping Agreements with a number of countries. In addition, its ships have undertaken coordinated patrolling and EEZ Surveillance on the request of neighbouring countries. Another element of ensuring safety of navigation in the IOR has been the hydrographic support provided to its partners to chart the waters of the region. This has been augmented with a large training and capacity building effort. In addition to the ASEAN and Indian Ocean Rim Association (IORA) mechanisms, the Indian Ocean Naval Symposium (IONS), of which India is a founding member, offers a broad-based platform for developing greater synergies with the Navies in the region. It also has well-established mechanisms like Regional Cooperation Agreement on Combating Piracy and Armed Robbery against Ships in Asia (ReCAAP) and Contact Group on Piracy off the Coast of Somalia (CGPCS), and anti-piracy patrols in the Gulf of Aden at the western extremity of this ocean.[14]

As of now only the Indian Navy is actively working towards security aspect of SAGAR in terms of joint exercises and patrolling under ambit of IONS but both IORA and IONS are entities with limitations and constraints in terms of their respective mandates, the political will of member countries etc. There is a need for devising a realistic structure within the framework of security and economic cooperation which is the linchpin of the concept of SAGAR. It has to be understood that approaches to security and economic development are interlinked and need to be integrated in the regional mindset. There are limits to the concept of multilateralism as it can be seen as a platform for few countries to pursue other interests or agenda, outside the equations of ongoing asymmetrical bilateral relationships. Disenchantment in regional bodies like ARF, European Union (EU) and international organisations like World Trade Organisation (WTO) and United Nations (UN) can be seen today.[15] Hence a governance system or network for the IOR is needed which can incorporate economic and security elements towards achievement of the goal of a secured and growing region.

Is SAGAR an Extension of India's Foreign Policy?

It is correct that through SAGAR India wants to apprise the world, specifically IOR countries of its thought process towards the Indian Ocean. Most of the time the MEA handles the propagation tasks of this policy. India's External Affairs Ministers generally mention this vision/policy in their talks on international platforms especially if matter of concern is IOR. But such an important, inclusive and integrated approach should be including maximum agencies for it to be successful.

The first key point of SAGAR speech of the PM, which is; "To do **everything** to safeguard India's mainland and islands and to **defend** its interests", underlines the priority of the vision. It means that India will safeguard its mainland and islands first, **at any cost** and will defend its interests. Since India's interests include safety of its SLOCs, it's EEZ and similar interests of its friendly oceanic neighbours, it will protect them as well.

Hence it is important to treat SAGAR as an extension of India's defence policy as well because for any maritime economy to survive, it is imperative to have a strong formidable maritime defence policy and structures/assets to support it. In the chapter of history, it became clear that there have been invasions and conquests of India from the land borders on previous occasions but they either led to transient political changes or to the foundation of new dynasties. The truth is that India never lost her independence till she lost command of the sea in the first decade of the sixteenth century.[16]

SAGAR can be seen as a declaration of national intent on the part of India to leverage its soft as well as hard power to shape the maritime order in the Indian Ocean, to meet the common challenges in the circumstances of deepening strategic flux.[17] This intent should not be taken into the direction of misinterpreting 'hard power' as hegemonic behaviour towards the territorially smaller nations who should get worried by India's defence posture into the Indian Ocean. India has never used its hard power in the IOR unless called upon and this policy will always remain so. India has the economic, military and technological capabilities to bolster her benign role and contribute significantly to the growth of the IOR. Given the situation of increasing '**strategic flux**', India has to improve its defence capabilities in all spheres of maritime '**Security**' not only for herself but for the peaceful continuation of '**Growth**' part of its SAGAR policy which is for the entire IOR.

NOTES

1. Sanjaya Baru, "India and the Indian Ocean Region: The New Geo-Economics", http://www.indiafoundation.in/india-and-the-indian-ocean-region-the-new-geo-economics-by-sanjaya-baru/ (Accessed 11 June 2018).
2. Valentina Romei and Reed John, "The Asian century is set to begin", https://www.ft.com/content/520cb6f6-2958-11e9-a5ab-ff8ef2b976c7 (Accessed 22 November 2019).
3. Harsh V. Pant, "India's challenge in the Indian ocean region", https://www.orfonline.org/expert-speak/indias-challenge-in-the-indian-ocean-region (Accessed 30 November 2018).
4. Arvind Gupta, "Strategic Overview of India's Maritime Challenges", Yogendra Kumar (ed.), *Whither Indian Ocean Maritime Order* (Delhi: K W Publisher, 2017), p. 221.
5. Shyam Saran, "A Strategic Perspective", Yogendra Kumar (ed.), *Whither Indian Ocean Maritime Order*, (Delhi: K W Publisher, 2017), p. 26.

6. Yogendra Kumar, "Introduction" (ed.) *Whither Indian Ocean Maritime Order*, (Delhi: K W Publisher, 2017), p. 10.
7. Narendra Modi, Indian PM, https://www.pmindia.gov.in/en/news_updates/text-of-the-pms-remarks-on-the-commissioning-of-coast-ship-barracuda/ (Accessed 12 March 2015).
8. Saran, op. cit., p. 23.
9. India Foundation is one of the leading think tanks which organises Indian Ocean Conference annually with different themes. It provides opportunity for the stake holders to deliberate on building institutional framework for managing the threats to regional peace and human security through the strengthening of multilateral cooperation based on mutual respect, equality and in line with international law. Till date four such conferences have taken place. This event is becoming popular worldwide because of the efficient conduct towards the aim of achieving concrete policy goals for the IOR.
10. Sushma Swaraj, https://www.mea.gov.in/Speeches-Statements.htm?dtl/ 30327/Remarks+by+External+Affairs+Minister+at+the+3rd+Indian+Ocean+Conference+Vietnam+August+27+2018 (Accessed 20 September 2018).
11. Ibid.
12. Ibid.
13. Ibid.
14. Ibid.
15. Leela Ponappa, "Overview of governance structures for the Indian Ocean Region", Yogendra Kumar (ed.), *Whither Indian Ocean Maritime Order* (Delhi: K W Publisher, 2017), p. 108.
16. K.M. Panikkar, *India and the Indian Ocean: an essay on the influence of sea power on Indian history* (London: George Allen and Unwin Ltd., 1945), p. 7.
17. Y. Kumar, op. cit., p. 11.

4

Strategic Flux in the IOR-Geopolitics or Ge-oil-itics

The IOR is home to most of the global hotspots of the world. Afghanistan, Pakistan, Iraq, Iran, Yemen, Somalia etc. are in constant state of turmoil. This has made the IOR focus of global attention, as major threat to global peace and tranquillity stems from this region and that's how after the Cold War era, US strategic focus has shifted from the Atlantic Ocean to the Indian Ocean.[1] One should acknowledge that the current security order in the IOR is underpinned mainly by the US through its Pacific Command (PACOM) and the Central Command (CENTCOM). The former lies to the east of Pakistan and since May 2018 is called the Indo Pacific Command (INDOPACOM) in recognition of the greater emphasis on the Indian subcontinent. The latter lies to the west of India. The US has a network of naval bases, a substantial fleet of naval vessels and security arrangements with several littoral countries. Post 1991, with the end of the Cold War, the US naval presence has been seen as relatively benign by India, which was not the case earlier. India opposed the extension of great power rivalry in the Indian Ocean along with opposing the use of Diego Garcia, belonging to Mauritius, as a naval base by the US.[2] Now, the Indian perception has changed and the US holds the largest number of naval exercises with India. The rise of China as a challenge to India's growth can be considered as one reason for bringing harmony in Indo-US relations.

India's geographical position, its peninsular shape and its island chains both to the east and the west, give it a uniquely advantageous position dominating the rapidly increasing maritime traffic linking the East coast of Africa to Australia, i.e., the Indian Ocean, the South China Sea and the Western Pacific. The Indian Navy is able to maintain a presence both in the eastern and western reaches of the Indian Ocean, but China too is emerging as a major naval and, indeed, as a maritime power. While China's current focus remains on the Yellow Sea, the Taiwan Straits and the South China Sea, its footprint across the Indian Ocean is steadily increasing. There is no doubt that the Indian and Chinese peripheries are going to overlap and intercept more and more as the two emerging powers build up their capabilities. But as the resource and power gap between India and China continues to widen, position of the Indian Navy is likely to erode in the coming years.[3] More recently, events like the development of the CPEC and the likely use of Gwadar as a Chinese naval base have also created fresh anxieties.[4]

The benign presence of the US in the last few decades has started changing character. India will face a major challenge in terms of security and economic aspects, if the US maritime strategy undergoes significant change. Presently, the US has a strong naval presence in the Gulf, which has traditionally safeguarded its energy security and that of its allies. However, its relative energy dependence on the countries of the region is likely to decline with the exploitation of shale oil and gas on the home front. The major fuel for growth for most of the Asian countries is oil. For the Asian countries, the supply of oil comes from West Asia—the Gulf region—while the major consumers, India, ASEAN, China, South Korea and Japan lie to the east. This oil, which fuels the growth of Asia, is transported via the Indian Ocean. The security of the production centres, as well as its transportation across the sea lanes, is thus vital for Asia. Oil, thus, is the most important governing factor of geopolitics these days and owing to its criticality, the word "**ge-oil-itics**"[5] needs to be introduced into common strategic parlance.

Although the exposure of the US to the Gulf is diminishing, in the security area, there has been no change from the past. During the last few years, discussions with the US have brought up the issue of India and China being free riders, depending upon American naval assets to protect their access to energy supplies from the Gulf without making any significant contribution themselves.[6] In the recent development on Afghanistan, the American president had mentioned that the US is "tired" of fighting everybody's war with terrorists in Afghanistan, which is some 7,000 miles away. He mentioned that it is also responsibility of Russia, Iran, Turkey, Pakistan and India as well.[7] The US has been planning draw-down of its troops after 19 years of long deployment in Afghanistan. It is clear that if there is a prospect of weakening of the US-led alliance system in the Middle East, it will invariably open the door for China to become even more assertive in the IOR. The Chinese objective of stabilizing a dominant position in the Asia Pacific region would be achieved and it could then turn its attention to the Indian Ocean.

Another convergence that can be seen gaining traction in the western IOR and increasing the strategic flux is the Russia-China-Iran trilateral engagement. The three countries held their first joint naval drill in the Indian Ocean in December 2019. Iranian naval chief, Rear Admiral Khanzadi, revealed that this was part of a broader naval collaboration between Iran and China, which also includes production of destroyers and submarines. The Foreign Minister of

Russia, Mr. Lavrov, had stated that the exercise was to train their forces in anti-piracy and anti-terrorism efforts. Irrespective of what is being said or not said publicly, the geopolitical underpinnings of the trilateral group are quite evident: all the three maintain a rather hostile relationship with the US and have shown increasing will to take cooperative actions in the security realm to illustrate their own capabilities. Beyond that, for countries like India, there are concerns about additional partners in this nexus, notably Pakistan, whose potential invitation has been the source of headlines, with Rear Admiral Khanzadi mentioning that Islamabad was welcome to participate. The coming together of China, Iran, and Russia itself is a major strategic development, and the addition of countries like Pakistan into the mix increases the stakes.[8]

Thus, if US power declines and there is exacerbation of conflict in India's western maritime region what would be the nature of challenges for India? Who will deal with these conflicts? Who will manage them? The Indian Navy is no substitute, not even in a modest way yet, for the US naval power there.[9] These questions stare at India against the backdrop of one fact-the Chinese naval footprint is expanding in the Indian Ocean. Chinese facilities at Gwadar and Djibouti may not qualify as full-fledged bases but they point in the direction of Chinese strategists thinking about China's emergence as a major power by having a global network of bases like the US. Under the BRI, China is acquiring port facilities across the IOR and in the world. The new dual-use ports it proposes to or has built in different littoral countries could also transform, in time, to naval facilities. Thus, the container terminal that China built at Colombo port also had Chinese submarines docking there in the recent past.[10] But before trying to answer the questions posed above, another aspect needs to be looked into which increases the intricacy of this flux.

Relations between the three major powers—India, China and the US—are evolving in complexity in the IOR theatre, heightened by the rise of India and China and a possible decline in US influence in the region. However, the perception of the US strategic decline warrants caution and should not be taken too literally. India recognises the US as a bulwark against Chinese expansion and assertiveness. In the broadest possible sense, while a rising China seeks to counter the dominance of the US and assert itself as the regional hegemon, an emerging India seeks, in turn, to act as a counter-balance against China. This is a situation that can serve the American interests well, as the two regional rivals are left to compete and take all or most of the risks involved. India's naval expansion is a case in point, as it adds additional weight to the US naval presence[11] and hence the increase in the naval exercises as well as naval defence deals with the US. This situation needs attentions of strategists to carefully draw the future path of self-reliant India with strategic autonomy in the region.

Gulf Problem: Trouble for the Asian Century?

A recent issue that is capable to derail the high hopes of making the 21st century an Asian century is not being talked about, specially by the stakeholders. The problem that should be raising the concerns of both, economists and strategic thinkers, is the danger to the **oil supply capability** of Saudi Arabia because of **stealthy air attacks** on its oil refineries on 14 September 2019. Strategists should worry about the small budget precision aerial weapons defeating heavily guarded installations of national importance (like Reliance refineries on India's west coast) and economists should calculate the dent in the GDP of India which is known to drop by 0.2 per cent with a USD 10 increase in per barrel oil price, according to Nomura[12] (a Japanese financial holding company), if this kind of situation occurs again. This attack by drones and cruise missiles has brought out the lethality and accuracy of air power while defeating another pillar of air power which is air defence (detection, identification, timely interception and destruction). This is a lesson which Indian policymakers, strategists and tacticians must learn fast and evaluate Indian air power preparedness for this kind of attack from land or maritime borders. Another thought to ponder over is that the Gulf region, being the main source of energy supplies for India, is also home to a six million strong diaspora and the sharpening sectarian conflict there has a backwash effect on a multi-religious and multi-cultural plural society like India's.[13]

The more worrisome fact is that none of the big Asian economies have vocally raised the level of concern on the deteriorating security environment in the Gulf. The shy response of the four Asian powers surfaces the absence of any strategy in Asia's capitals for dealing with security in the Persian Gulf if, or when, the Americans decide that they no longer want to get involved.[14] These four powers are India, China, Japan and South Korea. All of them depend heavily on oil imports from the Gulf for meeting their energy requirements. The last three claim to have sufficient cushion of strategic reserves; India's oil reserves are comparatively inadequate to cope with such turmoil in oil supply from the western Indian Ocean Region and would place in jeopardy, India's ambition of becoming a USD five trillion economy by 2024. In all, the future of geopolitics in the region will be controlled by "**ge-oil-itics**".

Non-Traditional Security Threats

Other than conventional rivalry for power balance and growing competition for force projection, there is a burgeoning concern over an array of non-traditional security threats in the IOR, especially threats to energy security. The Indian Ocean is undoubtedly critical to global trade and economic growth, as well as food and energy security but simultaneously, the world's most important trade corridor is also vulnerable to piracy and highly unpredictable potential acts of maritime terrorism. Maritime security can no longer be conceptualised only in terms of a composite of sea power and naval arms build-ups, island and maritime boundary disputes, navigational regimes, activities in EEZs, competition over resources, and the maintenance of law and order at sea, including the protection of SLOCs. The concept of human security also needs to be revisited with reference to some of the most pressing environmental issues such as land degradation, coastal erosion, access to fresh-water resources, exploitation of fishing stocks, climate change, illegal disposal of nuclear waste and urban expansion and deterioration.[15] Recurrent hurricanes in western IOR and devastating cyclones in eastern IOR (especially in the Bay of Bengal) along with frequent floods cause huge damage to life and material. South and south-eastern IOR are prone to tsunamis and earth quakes due to the vicinity of the Circum-Pacific belt which is also called the Ring of Fire.[16] The IOR is the locus of 70 per cent of the world's natural disasters, the majority of them having occurred in South-East Asia.[17] Displacement of millions of people due to conflicts or climate change has become a norm in western IOR. Maritime pollution, Weapons of Mass Destruction (WMD) proliferation and oil spillage are also factors that are creating instability in the IOR.[18] India itself has been one of the major victims of the huge nuclear-cum-missile proliferation conspiracy hatched by a China-North Korea-Pakistan axis and executed largely across the IOR.[19]

Of the three temperate oceans of the world, the Indian Ocean is probably the most problematic for security management. Despite this reality, the Ocean seems bereft of any collective maritime security arrangements. The lack of maritime security around the Horn of Africa, in particular, causes a great deal of concern as it not only threatens commerce, but also peace and regional stability, international trade and global energy flows. Current global realities have introduced a range of maritime security challenges in the IOR as the roles of non-state actors have direct and fundamental effects on the evolving situation.[20] This situation is causing emerging rivalries and the consequent militarisation is becoming evident. Even the countries that historically had little connection with the Indian Ocean are acquiring bases, developing their naval capabilities and updating their military doctrines to ensure a presence in the IOR. Since there is a lack of well-established security architecture in the region, the potential of military rivalry escalating cannot be ruled out.[21]

China's Military Overtures in the IOR–Threat to India?

Piracy off the Somali coast, the need for security of SLOCs in the Persia Gulf and requirements of naval escorts for the World Food Programme have contributed to an increase in the presence of several extra-regional navies in the IOR. The Americans, British and French have maintained a presence in the area and China is the latest entrant. The stated Chinese purpose is to conduct anti-piracy patrols, exploration and mining of sea-bed resources and participation in multilateral exercises but deploying submarines with support ships for anti-piracy drills does not add up. This brings out the Chinese aspiration to increase its strategic presence in the IOR. The other fact of concern is that instances of piracy in this region have reduced significantly but the tempo of Chinese anti-piracy patrols has increased with a new pattern being followed wherein the outgoing Task Force deploys to other areas such as the Mediterranean, Atlantic and Africa on being relieved. This, along with events in the South China Sea are signs that China is abandoning the traditional continental mind-set and extending its capabilities in the far seas as mentioned in the latest Chinese Defence White Paper.[22] China's ambitious BRI strategy, which has been categorised as an economic initiative by China, clearly has military intent. It is clear that their objective is to operate in a much broader area that includes the Indian Ocean.[23]

It is evident from recent developments and news in the open media that the unseen unrelenting paddling of an apparently calm China in the Indian Ocean has increased. In September 2019, seven Chinese (PLAN) warships were seen passing through the southern Indian Ocean including a 27,000 tonne-amphibious vessel. The Indian Navy's P-8I aircraft tracked their path.[24] They were reportedly heading towards the Gulf of Aden for routine deployment supposedly for anti-piracy drills as mentioned earlier. China has also launched its first amphibious assault ship capable of carrying helicopters.[25] Its Sharp Sword stealth Unmanned Combat Air Vehicle (UCAV) will be deployed aboard the Chinese carriers in the not-so-distant future.[26] Admiral John C. Aquilino, US Pacific Fleet Commander recently said that the increased weaponisation by China is a threat to free and like-minded nations and there is no doubt that those weapons are designed to continue to threaten the nations that they may have disputes with.[27]

With faster growth in technology and with deployment of its aerial assets on board these ships, whether manned or unmanned aircraft, China is planning to extend its air power umbrella far from its borders. It has started showing its might in the South China Sea already. And if today seven Chinese naval warships are seen in the Indian Ocean, witnessing a Carrier Battle Group (CBG) of PLAN with a full complement of air wing will not be too far in the future. India must assess its air power doctrine for readiness for such scenarios.

There are enough examples of initiatives taken by the Government in terms of improved IOR-oriented foreign and defence policies. The Indian Navy accordingly has been enhancing its blue water capabilities and becoming more global in terms of defence cooperation, expeditions and presence in all the waters as far as possible. But one must realise that may be this outgoing approach (in the IOR particularly) is not an **initiation** but a **reaction**. China did not venture much into the Indian Ocean until 2008. But it always had high value SLOCs all through the IOR. When Somalian piracy was at its peak and China wanted to secure bases in Eastern African countries, it started sending its naval ships with attack helicopters and Special Forces on board to assist in combating piracy.[28] And in 2008 itself India pioneered the IONS, starting a cooperation-based arrangement between navies towards the aim of having stable and secured international waters with freedom of navigation. This can be seen as an example of **reaction**.

China is building its third aircraft carrier and plans to make it sail soon in international waters. It has declared plans for building more carriers in the future. The Indian Navy has been raising the concern of having at least three carriers (if the required huge budgetary sanction is approved) to take care of the eastern and western areas of its operations (with one being under maintenance). This also can be seen as another **reaction**. Even the Army and the Air Force are bound to 'know thy enemy' and demand/maintain a certain level of preparedness in terms of assets, manpower and training. But defence procurements, demands or even policies should not be made based on reaction or competition. It may very well be the intention of an intelligent state to create challenges for its competitor (not in the war zone but even on the turf of economic progress) to draw out the resources of the contender and prevent it from playing the original game based on one's strength.

China and India can be seen in this context very well, where China does not want any other nation to be as prosperous in Asia (or may be in the whole world) and India definitely is a strong contender. India must realise that the Indian Ocean is her biggest strength and China would do (may be it is already doing) anything to prevent India to take a leading role in the IOR. In terms of having a bigger navy, army or air force, India cannot match China and that really is not the intention either. Vice Admiral Satish Soni (Retd) was candid enough in a seminar to say that in terms of naval assets, India's numbers are woefully inadequate to meet the growing requirements and there are capability gaps in terms of hardware. The Indian Navy may be the dominant navy in the region but resources are somewhat stretched to suit the role of a net security provider especially in light of growing Chinese presence.[29] Here one must understand that India as a nation can assume the role of a net security provider in the region while coordinating its assets well instead of a single service trying to accomplish the humongous task. As the Chief of Air Staff of the IAF mentioned in an interview that over the years the IAF has modernised and enhanced its capabilities fairly well and it can play a significant role in being a part of net security provider capability of the country.[30]

As far as China building up its carrier fleet is concerned, one must understand that it is taking its war front away from home. Sounds familiar? Yes, because America has done the same and said the same for so many years. China wants to assure some face of strength to be present in the oceans, be it the Pacific, Indian or Indo-Pacific. China claims most of the SCS, marking its maps with the infamous "9-dash line," overlapping with several other Southeast Asian nations. China has not only violated almost everyone else's waters, but also constructed artificial islands there.[31] After making its island chains secured frontiers for guarding the mainland, it has started expeditions in other oceans because of the distances from its stakes. These stakes are investments in countries that it has trade relations with as well as lanes of energy supply through the Indian Ocean. India is in a position to secure its stakes in the IOR from home itself given the geographic location of the mainland as well as the island chains. Hence, China's naval expansion in terms of carriers, submarines, etc. is driven by its expansionist strategy. As it is, carrier-to-carrier battle is in the history books of World War II[32] and does not seem possible in the present scenario where carriers are very expensive and anti-carrier weapons are cheaper unless one has naval assets to create supremacy in areas away from home (like the US). "It is obvious that under these circumstances, you must make sure that you have an air force above you."[33]

Is China the Main Factor Behind Indian Defence Philosophy?

Before the independence of India, the Indian Ocean was called a British lake. The naval supremacy of the Royal Navy was uncontested and India did not have to worry about the IOR (rather, that mind set was missing). Post World War II, it was the US Navy that started showing its presence which grew gradually and even impacted Indian course of actions (1971 war with Pakistan) with its presence in the Bay of Bengal. The Cold War saw the tussle between the US and Russia in the region. The British, French and Australian territories (dependencies) are still manned by their respective administrations with adequate presence on the ground. And hence the Chinese presence can be seen as a result of the common question by the big powers, "Why not us?"

The point here is that the Indian Ocean i.e. the Hind Mahasagar (which was named after India centuries ago) has been a historical source of prosperity and peaceful growth and India has always played a major role in this process. In other words, India has long been a quiet spectator of events in its own arena without having the strong credibility of a performer. In the last decade India has shown some initiative to take up the responsibility of an important stakeholder in the IOR in terms of naval diplomacy. The Indian Prime Minister's outreach towards the IOR and ASEAN countries has brought in freshness in India's international relations. The visits by the President, the PM and other dignitaries to IOR countries and inviting the heads of state from those countries to India have rejuvenated the centuries-old ties to a certain extent. In terms of defence cooperation also, the efforts are seen to be on the rise as brought out by the data on international exercises conducted by the Indian Defence Forces since 2012 (see chapter 5). But all this progress seems to have been triggered (in a way) in response to the act of other countries which have been or started, showing an indubitable intent of dominating the IOR. Indian philosophy should not be competition-oriented but cooperation-oriented and all smaller countries should be included in active and continuous engagements on all fronts under this philosophy.

Today, despite the way the US looks at China and its growth (economic and military) and opposes its hegemonic attitude, it is not necessary for India to wear the same lens but to have its own strategy in place. For that matter, can India completely rely on the US Navy for assisting it in any naval confrontation with Pakistan in the future? On the other hand, can India be sure that China will assist Pakistan in such a scenario? Improving the defence posture in the IOR by enhancing air power practices to augment the naval presence should be done with the aim of creating a strategic envelope around the Indian maritime domain as well as other friendly IOR nations (so that any extra regional entity must not think of any misadventure). Because India respects and follows the principle of freedom of navigation and abides by the United Nations Convention on the Law of the Sea (UNCLOS), a responsible posture by India will ensure that there is no violation of rules and no challenges to sovereignty, assets, SLOCs and peace of any of the IOR countries.

The ever placid Indian Ocean waters are becoming choppy. The challenges lie in managing the geostrategic shift in favour of the regional countries and also in preventing the disturbances from neighbouring seas to shift into the Indian Ocean. The appearance of the Chinese Navy in the Indian Ocean is not threatening (at the moment at least) but it is unsettling, not only for India or the littoral countries of IOR, but even for the US, Japan, Australia, France and other countries having their interest in the IOR. The Prime Minister, while recognising this extra-regional interest, placed the primary responsibility for peace, stability and prosperity on the countries of the Indian Ocean littoral. All countries including extra-regional economies support India's strategic role in the IOR and are willing to cooperate in enhancing it. SAGAR can be a strategic framework for nurturing a foundation of the shared values and a common vision which is consonant with India's millennial civilisational links.[34] It is time that India should use the SAGAR policy to refine its maritime defence policy as a regional defence policy with maximum IOR countries on board and assume the role of a regional security provider.

NOTES

1. Alok Bansal, "Challenges to IOR Security-Indian Strategy and Options to Counter Them", Bhaskar C Uday and Agnihotri Kamlesh K (Eds.), *Security Challenges Along the Indian Ocean Littoral-Indian and US Perspectives*, (New Delhi: Matrix Publishers, 2011), p. 107.
2. Shyam Saran, "A Strategic Perspective", Yogendra Kumar (ed.), *Whither Indian Ocean Maritime Order*, (Delhi, K W Publisher, 2017), p. 27.
3. Ibid., p. 25.
4. Ibid., p. 28.
5. Vikas Kalyani, The word 'Ge-oil-itics' was first used by the author in an article on the imbroglio in the gulf after attack on oil refineries in Saudi Arab. 'GULF IMBROGLIO: CHANGING GE-OIL-ITICS', https://indiafoundation.in/articles-and-papers/gulf-imbroglio-changing-ge-oil-itics/.
6. Ibid.
7. Press Trust of India, "India and other countries must fight terrorists in Afghanistan at some point, says Donald Trump", *First Post*, 22 August 2019.
8. Rajeswari Pillai, "What's in the Growing Russia-China-Iran Trilateral Convergence?", *The Diplomat*, 23 December 2019.
9. Saran, op. cit., p. 29.

10. Ibid., p. 30.
11. Denis Venter, "India and Africa: Maritime Security and India's Strategic Interests in the Western Indian Ocean", in Iain Walker, Manuel João Ramos and Preben Kaarsholm (eds), *Fluid Networks and Hegemonic Powers in the Western Indian Ocean*, 2017, pp. 131-167, e-version at https://books.openedition.org/cei/469?lang=en (Accessed 12 November 2019).
12. Hasan Alhasan, https://www.asiatimes.com/2019/09/opinion/big-asian-players-too-quiet-on-threats-to-gulf-security (Accessed 01 October 2019).
13. Saran, op. cit., p. 28.
14. Hasan, op. cit.
15. Venter, op. cit.
16. Ring of Fire, https://www.nationalgeographic.org/encyclopedia/ring-fire/ (Accessed 10 November 2019).
17. Shishir Upadhyaya, "Mechanisms for Cooperation and Confidence Building in the IOR-an Indian Perspective", Bhaskar C Uday and Agnihotri Kamlesh K (Eds.), *Security Challenges Along the Indian Ocean Littoral-Indian and US Perspectives*, (New Delhi: Matrix Publishers, 2011), p. 125.
18. Arvind Gupta, "Strategic Overview of India's Maritime Challenges", Yogendra Kumar (ed.), *Whither Indian Ocean Maritime Order* (Delhi: K W Publisher, 2017), p.221.
19. Arun Prakash, "Introduction", Bhaskar C Uday and Agnihotri Kamlesh K (Eds.), *Security Challenges Along the Indian Ocean Littoral-Indian and US Perspectives* (New Delhi: Matrix Publishers, 2011), p. 3.
20. Venter, op. cit.
21. Gupta, op. cit.
22. Satish Soni, "Traditional Security Challenges to the Maritime Order in the IOR", Yogendra Kumar (ed.), *Whither Indian Ocean Maritime Order* (Delhi: K W Publisher, 2017), p. 79.
23. Dinakar Peri, "Weaponisation by China a threat: U.S. Admiral", *The Hindu*, 30 October 2019.
24. Manjeet Negi, https://www.indiatoday.in/india/story/navy-tracks-down-chinese-nuclear-submarines-operating-close-to-indian-waters-1599555-2019-09-16 (Accessed 29 September 2019).
25. Sutirtho Patranobis, https://www.hindustantimes.com/world-news/china-launches-its-first-amphibious-warship/story-j5V2P6feCuHFGJYIgHlRTO.html (Accessed 29 September 2019).
26. Rogoway Tyler, https://www.thedrive.com/the-war-zone/30020/china-deploying-a-weaponless-stealth-drone-on-its-carriers-makes-perfect-sense (Accessed 29 September 2019).
27. Peri, op. cit.
28. Mark McDonald, "China Sends Naval Task Force on Anti-Piracy Mission", *The New York Times*, 26 December 2008.
29. Soni, op. cit., p. 86.
30. Air Chief Marshal RKS Bhadauria, Chief of Air Staff, IAF, Raisina Dialogue 2020, 15 January 2020.

31. Karim Raslan, "OPINION: In the South China Sea, ASEAN needs to stand united", 14 January 2020, https://news.abs-cbn.com/blogs/opinions/01/14/20/opinion-in-the-south-china-sea-asean-needs-to-stand-united (Accessed 20 January 2020).
32. Warfare History Network, "FACT: The Last Aircraft Carrier vs. Aircraft Carrier Battle Was Decades Ago: A World War II showdown", 17 November 2019, https://nationalinterest.org/blog/buzz/fact-last-aircraft-carrier-vs-aircraft-carrier-battle-was-decades-ago-96846 (Accessed 12 December 2019).
33. Jasjit Singh, "National Security and Aircraft Carriers", in Bhaskar C Uday, Upadhyaya Shishir (Eds.), *The Aircraft Carrier in the 21st Century* (New Delhi: KW Publishers, 2011), p. 135
34. Yogendra Kumar, "Introduction" (ed.), *Whither Indian Ocean Maritime Order*, (Delhi: K W Publisher, 2017), p. 10.

5

Indian Defence Forces in the IOR

It has been discussed adequately that through vision SAGAR, India is capable of assuming the responsibility of regional security provider in the IOR and that will ensure peaceful growth of the region. While the vision of India's national leadership is becoming ocean-centric, the gap between national intention and action however, needs to be bridged, or in other words, the gap between vision and policy needs to be closed. SAGAR can be considered as a vision or base of India's policy of outreach in the IOR which should be looked at as a source of rejuvenation of doctrines and ultimately, creation of new strategic thought process.

By now it should be reasonably clear that the Indian Ocean has been and will remain a source of potential growth, not only for India but for all the IOR countries and many beyond the region. India, in partnership with other oceanic neighbours, bears the task to keep the IOR as safe and secure as possible. As the leading nation in the IOR, India has assumed the responsibility of assuring preservation of the rule based order and free movement of goods and vessels in the global commons.

History has shown that the old IOR ecosystem was peaceful and prosperous with all participants following rules of harmony and co-existence. Gradually, the spread of western powers into Indian Ocean and subsequent phase of trade wars and colonisation turned the 'mutual growth of all' into 'richness of few' (powerful) nations. That is when India also faced the downfall of prosperity under British Raj because of weaker maritime defence and lack of national vision. Importance of strong maritime frontiers is thus a prerequisite for development. Today, India has a national vision and stronger maritime defence; however, the process needs constant review with assessment of future growth through secured Indian Ocean for enhancing current capabilities. A strong maritime nation would require greater air power prowess as an essential component, along with the traditional components of robust maritime trade and resilient blue water navy. The fall of Singapore in front of Japanese air power in World War 2, capturing of the Andaman Islands because of high degree of control of air by Japanese forces, India using air power over land as well as in the Bay of Bengal to weaken East Pakistan's position in 1971 war and present control by the US over situations in Gulf and adjoining ocean with extended air arm are amongst such examples which support the fact that to control the seas, control of air is also essential.

Pannikar had assessed the need for a holistic maritime defence policy as far back as in 1945 when he postulated, "the fact can never be overlooked that the economic life of India will be completely at the mercy of the power which controls the seas. Also the security of India could be perpetually threatened, for it is but seldom that fortified land posts held under the cover of naval guns by a power having the command of the seas have been successfully attacked from the land side. With the open coastline of over 2,000 miles, there will clearly be no safety for India, if the Indian Ocean ceases to be a protected sea." He went on to state, "the indivisible sea is no longer indivisible. **Air power has introduced a new factor in the control of seas.**"[1]

In order to have a protected Indian Ocean, as the historian also mentioned, an understanding followed by a system having security as priority is needed for the IOR. Security forces of all the nations have to be exposed to think beyond their local area of responsibility and realise the importance of a secured Indian Ocean. The Indian Defence Forces have been responding to situations threatening the stability of the region and will continue to do so. However, there is scope for the countries of the region to be proactive and be more cohesive towards the joint responsibility of protecting the IOR. To appreciate the role of Indian forces for providing assistance to IOR countries in case of any threat (conventional or non-conventional) and succour in case of natural or man-made disasters, it is better to have a look at past examples of Indian forces' contribution in furthering peace, trust and stability in IOR and beyond. These can be also termed as confidence building capabilities of Indian defence forces.

Humanitarian Assistance and Disaster Relief (HADR) Operations

As one of the most devoted, qualified, and contemporary militaries in the world, the Indian Military can respond to any catastrophic situation with all their strength. It is due to their technical competence, trained manpower, and logistical capabilities that they are always ready to rapidly undertake any kind of disaster-related rescue and relief operation. In the 21st century, India and other IOR countries have been facing increased frequency and magnitude of natural disasters. Post 2004 tsunami, the Indian defence forces, coordinated by the Integrated Defence Staff (IDS), efficiently handled relief, rescue and evacuation work under Operation Sea Wave at local front, while extending aid to Sri Lanka and Maldives under

Operation Rainbow and Operation Castor respectively.[2] The Indian help even reached as far as Indonesia for limited duration of 90 days. India has been the "first responder" to calls of assistance e.g. providing relief supplies and medical assistance to flood ravaged people of Sri Lanka in June 2017, rescuing Bangladeshis swept off the coast due to cyclone Mora, alleviating the acute drinking water crisis in Maldives in 2014 when the IAF airlifted 1,000 tonnes of fresh water to Male. India also dispatched cyclone relief materials to Fiji in February 2016.[3]

Peace Keeping Operations

For discharging its obligations towards global peace and as a stakeholder in the international security architecture, India is one of the largest contributors of troops for UN mandated peace keeping operations. India has been at the vanguard of peace keeping right from 1950 when it supplied medical personnel and troops to the UN Repatriation Commission in Korea. Since then, India has participated in 49 missions, sending more than 208,000 troops. 156 Indians have sacrificed their lives in these peace keeping operations, the largest sacrifice by any troop-contributing nation[4] Undoubtedly the professional excellence of the Indian troops has won universal admiration.

Combating Piracy at Sea

Successful patrolling through 'Presence-cum-Surveillance' missions in piracy infested waters in the Gulf of Aden by the Indian Navy since October 2008 and thwarting of attacks by the pirates has facilitated in protecting India's sea borne trade and instilled confidence in its sea faring community. The decision to deploy warships in the region was taken by the government after Somali pirates hijacked Japanese-owned chemical tanker MV Stolt Valor on 15 September 2008.[5] Due to increased cooperation among different navies, with task forces from the US and the European Union also operating in the region, the incidents of piracy have gone down. India too is contributing its might to deter piracy by patrolling the normal route followed by Indian flagged ships during passage from Salalah (Oman) to Aden (Yemen) as well as by coordinating with other foreign navies.[6] There are MoUs in place with Sri Lanka and Maldives for joint/coordinated patrolling. Sharing of MDA is also in practice through Information Management and Analysis Centre (IMAC) which is an initiative of the Indian Navy. This control centre obtains feeds from a range of space-based and terrestrial sensors as well as other sources to track fishing boats and commercial vessels near India's coast and in the vast maritime domain beyond. In 2018, India's defence minister invited regional countries as well as other stake holders to join an international version of IMAC called the Information Fusion Centre for the Indian Ocean Region (IFC-IOR).[7]

Non-Combatant Evacuation (NCE) Operations

Non-Combatant Evacuation Operations are missions to evacuate Indian citizens, and sometimes citizens of other friendly countries, from unstable, potentially dangerous and conflict afflicted areas. The successful evacuation of 111,711 Indian citizens from Iraq, Kuwait and Jordan in 1990 by operating 488 refugee flights over a period of 59 days by Air India and the IAF has been a world record.[8] Similar missions were carried out from Lebanon in 2006 (Operation Sukoon), Libya in 2011 and Yemen (Operation Rahat) in April 2015. Operation Rahat, involved the deployment of C-17 aircraft of the IAF and Air India's aircraft to airlift thousands of people in distress. The operation also had Indian ships involved in the safe evacuation of Indian expatriates and foreign nationals from Yemen. The Indian government successfully evacuated almost 5,000 Indians and nearly 1,000 citizens from 41 other countries.[9]

In order to carry out all these international missions, it is essential that Indian defence forces are on the same frequency with other forces of the region and are maintaining goodwill with them. Defence cooperation and international exercises are very crucial in achieving the desired level of efficiency during real time operations.

Defence Cooperation

Towards usage of a nation's 'soft power' as part of defence diplomacy, role of international defence cooperation in gainfully supplementing diplomatic initiatives, is increasingly being acknowledged the world over.[10] Indian government has increased the engagement in this field bilaterally and multilaterally in the last few years, focused on capacity building and capability enhancement of neighbouring countries in the IOR. Towards a secured IOR, Naval cooperation is one of the most promising areas of military-to-military cooperation by virtue of its capacity to safeguard critical SLOCs and establish linkages through maritime bonding,[11] but number of such engagements are inadequate and irregular with IOR countries and participation of the IAF is minuscule. International exercises are a good medium to build mutual trust and enhance cooperation. Few charts and figures are taken from the article 'International Military Exercises: An Indian Perspective' by Group Captain KK Khera in IDSA's Journal of Defence Studies to understand the trend of international engagements by Indian forces.

Exercises Conducted with Other Countries

According to the Annual Reports of the Ministry of Defence (MoD), Government of India, from 2011–12 to 2015–16, the Indian defence forces have been engaging a number of countries through regular international military exercises in India and abroad. The share of various countries in these international exercises by Indian defence forces is given on the world's map (in Fig. 5.1). Partner countries for international military exercises that India conducted or was part of

Fig. 5.1: Partner Countries for International Military Exercises with India from 2012 onwards

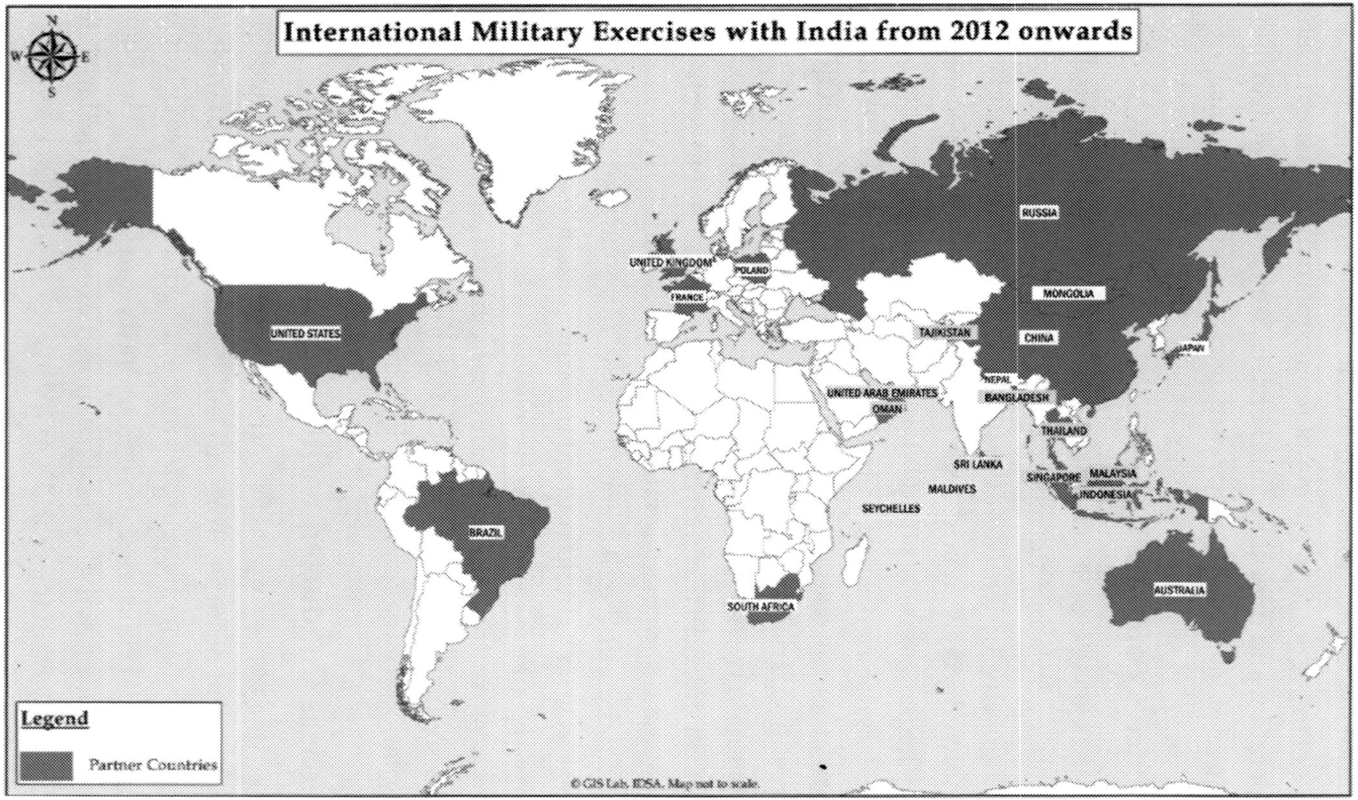

Note: Countries participated in joint military exercises with India from 2012 onwards.
Source: Ministry of Defence, *Annual Reports* (2011–12 to 2015–16).

from 2012 onwards are depicted in Fig. 5.2. **Singapore** and **the US** head the list with 12 engagements since 2012 amongst the list of 23 countries in this category.[12]

Fig. 5.2: Depicting Countries Engaged by the Indian Defence Forces in Military Exercises since 2012 along with Frequency

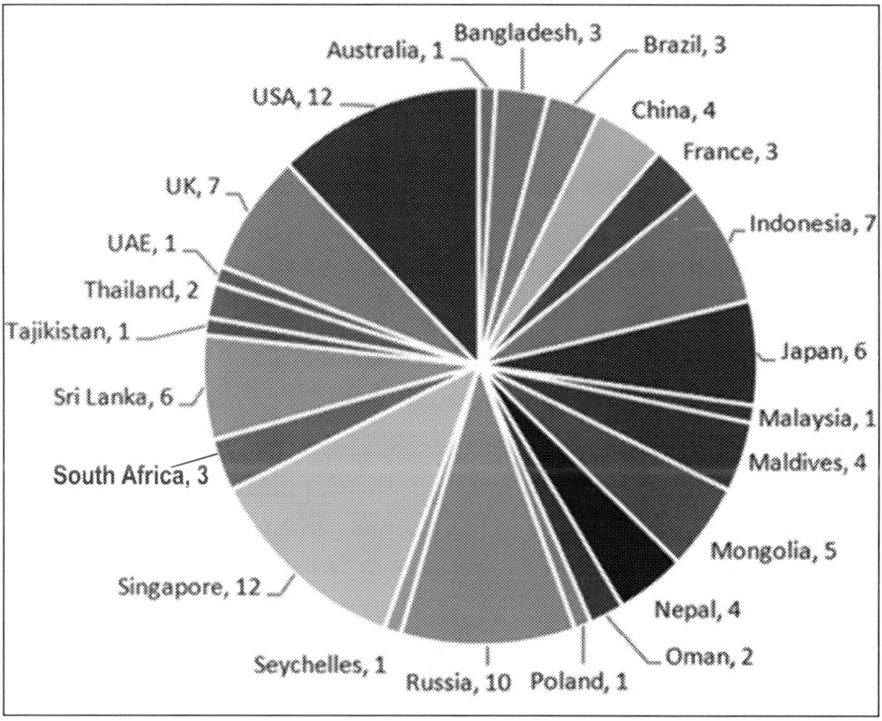

Source: Ministry of Defence, Annual Reports (2011–12 to 2015–16).

It is evident that engagement with most of the neighbouring IOR countries has been meagre. In such exercise, Sri Lanka tops the list with only 6 exercises (out of 93), that too four by the Indian Army, two by the Indian Navy and **none by the IAF**.

The Indian Army has expanded the engagement envelope for international exercises focused on various domains of warfare that are decided mutually with the partner countries. The Indian Navy regularly conducts exercises with various foreign navies at both bilateral and multilateral levels in three participative formats:

passage exercises (PASSEX); occasional exercises; and institutionalised exercises. PASSEX are conducted when ships of either navy pass by other's coast, with the duration and complexity determined as per the operational considerations and degree of interoperability attained. Occasional exercises are those periodic international exercises on maritime security or HADR in which the Indian Navy participates occasionally on the side-lines of overseas deployments. Institutionalised exercises are conducted on regular basis with partner countries in areas of maritime interest, wherein the scope and content are progressively enhanced for traditional and non-traditional maritime challenges.[13] The largest number of exercises carried out by the US Navy is with India. However, Indo-US naval cooperation is mainly focused in the eastern IOR. The reason may be increased interest of the US in its INDO-PACOM area (thus the name-change of erstwhile PACOM). There is little activity on the western reaches of IOR where the security challenges for India are more significant given the dependence on oil supply from the Gulf, the large Indian diaspora in Middle East and now Gwadar port being highlight of the CPEC project.[14] The Indian Navy has also spearheaded conduct of first IONS meet in 2008 in India, which provided an ideal forum for maritime confidence building measures and capacity building to address various asymmetric and transnational threats in the IOR.[15] Presently, there are 35 countries as members of this forum including South Asian, West Asian, East African, South East Asian and Australian littorals. The Indian Navy also began holding friendly naval exercises codenamed Milan, out of Port Blair in the Andamans in 1995. Milan began with only five countries in that year but in 2018 there were 16 navies of Indian Ocean countries.[16]

International Exercises by the IAF

On the other hand, the IAF has come a long way after its first international exercise with participation of fighter aircraft, Exercise

Garuda-I with France, in India in 2003. Other exercises in various formats have also been conducted since then. These include: Exercise Cope Thunder, later renamed Exercise Red Flag-Alaska, Exercise Red Flag-Nellis and Exercise Cope India (with the US since 2004); Exercise Golden Eagle (with South Africa since 2004); Exercise Indra Dhanush (with the UK since 2006); Exercise Desert Eagle (with the UAE since 2008); Exercise Eastern Bridge (with Oman since 2008); Joint Military Training (JMT) (with Singapore since 2007); Exercise AviaIndra (with Russia since 2014); and Exercise Siam Bharat (with Thailand since 2015).[17]

Statistics

The pattern of engagement of Indian Military with other militaries is depicted in Fig. 5.3, which indicates 55 per cent share for Asian countries in international military exercises. Africa's share, at just 4 per cent, is considerably lower than its strategic significance (especially East African countries which come in IOR).[18]

Fig. 5.3: Continent-wise Share of International Military Exercises by India since 2012

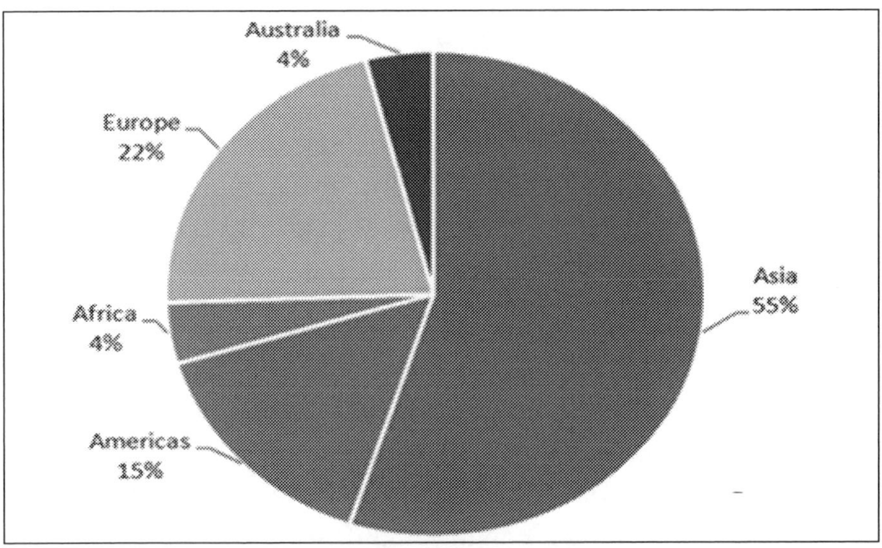

Source: Ministry of Defence, *Annual Reports* (2011–12 to 2015–16).

In this context, the largest engagement is by the Indian Army (Fig. 5.4). The probable reason for this is the multiple roles the army needs to play (being the largest of the three services), in the prevailing environment in counter-insurgency/anti-terrorist operations, special operations and disaster relief.

Fig. 5.4: Share of Individual Services in India's Participation in International Military Exercises since 2012

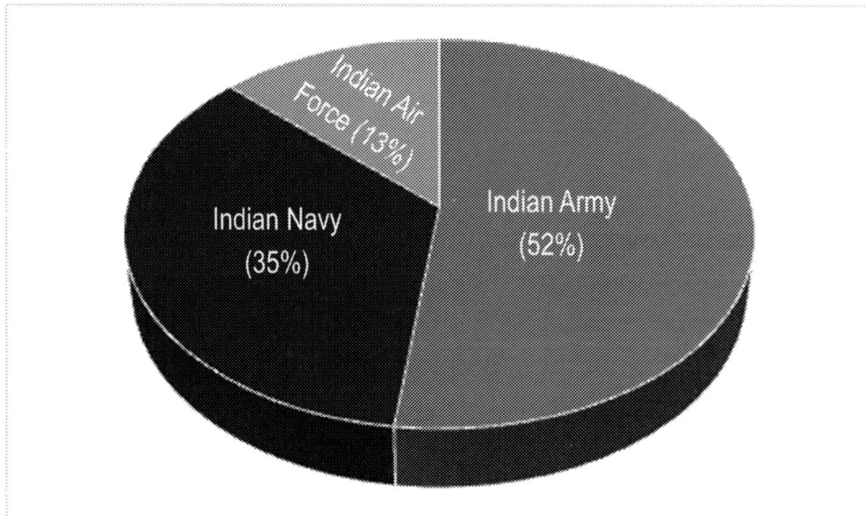

Source: Ministry of Defence, *Annual Reports* (2011–12 to 2015–16) and Year End Review-2017, 2018: Ministry of Defence, Press Information Bureau

The Indian Army is, on an average, participating in 10 international exercises annually; the Indian Navy in seven exercises besides PASSEX; and the IAF averages three (only) per year as depicted in Fig. 5.5.[19]

Less Exposure to the IAF?

From the graph above it is clear that the IAF has got least exposure to the international engagements overall and very less in the IOR. There are many reasons for the low number of exercises involving the air force. An exercise involving aircraft, especially fighter aircraft,

Fig. 5.5: Depicting Year-wise Share of the Three Services in Participation in International Military Exercises since 2012

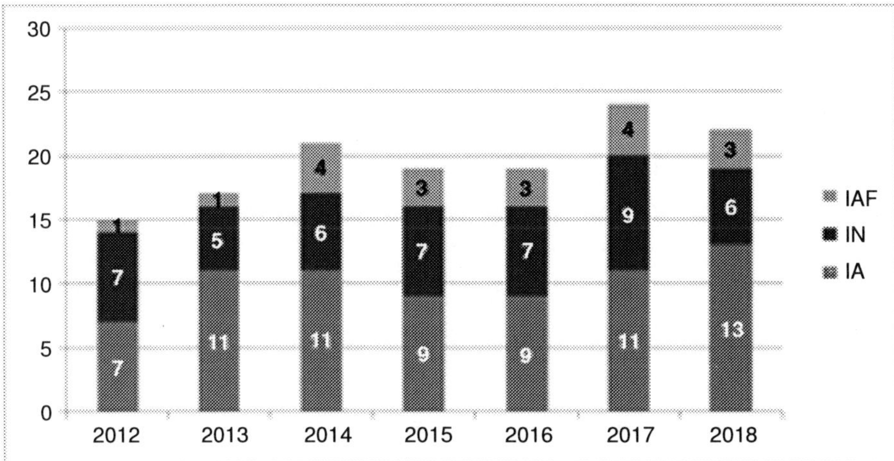

Source: Ministry of Defence, *Annual Reports* (2011–12 to 2015–16) and Year End Review-2017, 2018: Ministry of Defence, Press Information Bureau

is considered to be complex because of its impact and optics. Exercises with fighter aircraft are intricate and necessitate a large coordination effort involving the participant country and the large number of countries enroute to ensure diplomatic clearance for over flight. Apprehensions in certain sectors/countries for permitting fighter aircraft to approach from certain directions and restricted aerial refuelling corridors make planning for international military exercises with fighter aircraft a challenging task. Another area of concern, while planning an international exercise with fighter aircraft is the availability of suitable diversions enroute to the planned destination. A large capability differential between participants restricts usefulness of an aerial combat exercise. This restricts the number of air forces which can engage in military exercises with combat operations as a basic theme. However, for HADR operations or application of air power in sub-conventional operations, such near parity is not essential.[20]

Flexibility in Cooperation

All the points given above are justifying enough for the reason why the frequency of the IAF is least for such engagements. All the main exercises are with bigger and far off countries like US, UK, Russia and even Australia. These big exercises generally involve a lot of equipment and manpower. Fighter aircraft with FRA, their staging halts, transport support for carrying technical equipment and ground crew, their diplomatic clearances and stay, all this costs a lot of money. All this can be done at reduced cost when oceanic neighbours are engaged with proper planning. All the constraints, in context of the IOR, then should appear to be easily workable. Most of the constraints are regarding fighter aircraft. The IAF has fourth largest inventory in the world and the biggest in the IOR. It need not use fighter aircraft for all engagements with friendly IOR countries (as far as the point of capability differential is concerned) unless the other country so desires. Even if there is an agreement or requirement for fighter aircraft to be used, many countries are within such distance that excessive planning regarding aerial refuelling will not be required, especially if the island chains of Andaman-Nicobar or Lakshadweep are used as the launch pads. There is no need to have a large scale combat exercise every time.

The IAF can also conduct Passex in which only one or two aircraft shall do a cross country ferry to a neighbouring country's airbase within range, interact with the hosting base staff and return same day or next day after turn round servicing. Transport aircraft fleet of the IAF is well equipped with modern aircraft capable of covering large distances across the IOR with equipment and materials required by the IOR neighbours in time of need or with troops when engaged in exercises. For example, the Cocos Keeling Island and Christmas Island (Australian dependencies) have 8000 and 7000 feet runways respectively and are fit for operations by all kinds of aircraft. They are located south of Indonesia and almost midway between

India and Australia (Fig. 5.6). Distance between Port Blair and these islands is 2675 km and 2833 km respectively which can be covered easily by aircraft launched from Port Blair. Similarly, A&N Islands of India can be used for same purpose.[21] The IAF must engage other Eastern IOR countries on regular basis. Exercises with Thailand and Singapore are frequent but the IAF has not interacted with Japanese Defence Forces in an air exercise yet which must be looked into. Indonesia, Vietnam, Malaysia and Taiwan are also important countries to have good frequency with. Basically, healthy engagement with ASEAN countries in eastern Indian Ocean should be the focus. Other IOR nations with which India has good relations and must build stronger ties are Mauritius, Seychelles, Mozambique,

Fig. 5.6: Cocos Keeling Island and Christmas Island (Australian dependencies)

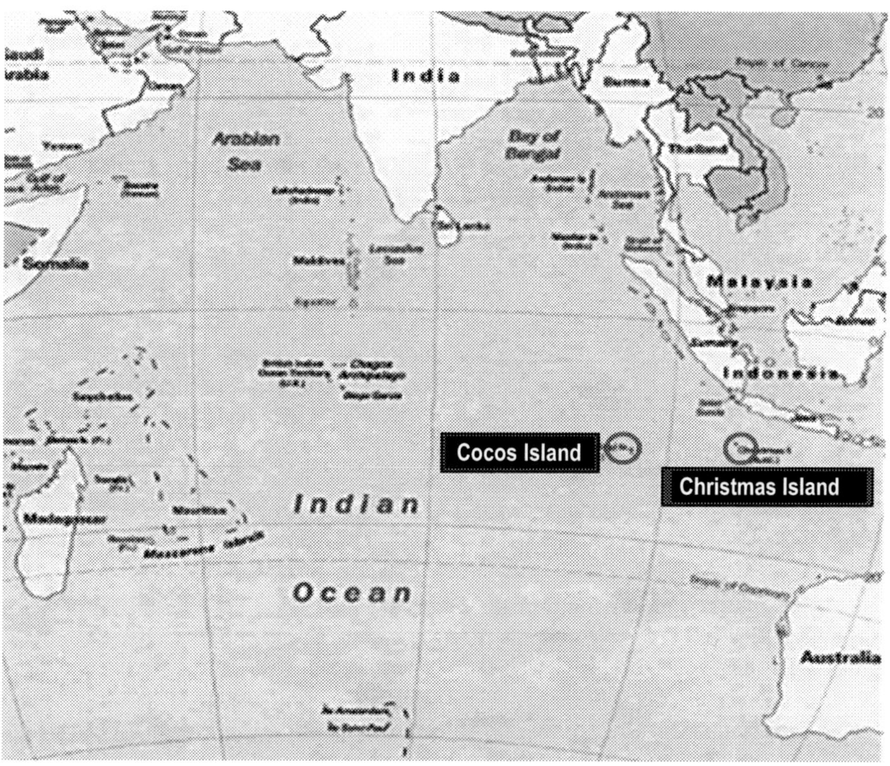

Madagascar, Oman, Qatar, Maldives and Sri Lanka.[22] The mutual respect with these countries is of good order and their geographical location in western and north western Indian Ocean is significant given the strategic concerns mentioned before. Mutual cooperation oriented air operations from these places for training, maritime surveillance or transit will dissuade the threats to the security of the region.

All these practices will help in area familiarisation, air space familiarisation and procedure familiarisation and thus enhance air force to air force relations. The technical capabilities possessed by the IAF in terms of surveillance aircraft and radars as well as integrated communication systems are other force multipliers to enhance overall security dimension in the IOR. It is therefore important to look into the need of increased engagement of the IAF within the IOR.

NOTES

1. K.M. Panikkar, *India and the Indian Ocean: an essay on the influence of sea power on Indian history* (London: George Allen and Unwin Ltd., 1945), pp. 14-15.
2. "Relief ops 'Castor' and 'Rainbow' launched", *Times of India*, 27 December 2004, https://timesofindia.indiatimes.com/india/Relief-ops-Castor-and-Rainbow-launched/articleshow/972740.cms
3. Sushma Swaraj, https://indiafoundation.in/articles-and-papers/sagar-indias-vision-for-the-indian-ocean-region/ (Accessed 15 May 2019).
4. Pallav Agarwal, "70 years of peacekeeping: Achievements, challenges and need for reforms", 10 December 2018, https://www.orfonline.org/expert-speak/70-years-of-peacekeeping-achievements-challenges-and-need-for-reforms-46135/ (Accessed 10 October 2019).
5. "India calls for UN force to deter pirates off Somalia", *Times of India*, 13 November 2008, https://timesofindia.indiatimes.com/india/India-calls-for-UN-force-to-deter-pirates-off-Somalia/articleshow/3710102.cms.
6. Ibid.
7. Ajai Shukla, "Navy invites neighbours to join Indian Ocean monitoring facility in Gurgaon", *Business Standard*, 23 December 2018, https://www.business-standard.com/article/defence/navy-invites-neighbours-to-join-indian-ocean-monitoring-facility-in-gurgaon-118122300007_1.html. (Accessed 11 October 2019).
8. IDSA, "Net Security Provider: India's Out-of-Area Contingency Operation",

https://idsa.in/system/files/book/book_NetSecurityProvider.pdf, p. 27. (Accessed 10 October 2019).

9. Constantino Xavier, "Getting Back Home, Safely", https://carnegieindia.org/2017/02/08/getting-back-home-safely-pub-67974. (Accessed 12 November 2019).
10. Arvind Dutta, "Role of India's Defence Cooperation Initiatives in Meeting the Foreign Policy Goals", *Journal of Defence Studies*, Vol. 3, No. 3 (July 2009), p. 31.
11. Ibid., p. 35.
12. Kishore Kumar Khera, "International Military Exercises: An Indian Perspective", *Journal of Defence Studies*, Vol. 11, No. 3 (July-September 2017), pp. 17-40.
13. Ibid., p. 25.
14. Shyam Saran, "A Strategic Perspective", Yogendra Kumar (ed.), *Whither Indian Ocean Maritime Order* (Delhi: K W Publisher, 2017), p. 28.
15. Dutta, op. cit.
16. Saran, op. cit., p. 27.
17. Khera, op. cit., p. 26.
18. Ibid.
19. Ibid., p. 28.
20. Ibid., p. 29.
21. Vikas Kalyani, "Australian view for Indian Ocean: significance for India", *Salute*, Vol. 11, No 11 (July-August 2019), pp. 20-21. There are many island territories in the IOR which belong to different countries. Like Reunion Island of France, Diego Garcia of US, Christian and Coco Keeling Island of Australia etc. Most of these countries have good relationship with India because they understand and respect India's geo-political importance in the region and wish to develop strategic partnership with India. France and US have signed many defence agreements regarding logistics support to each other's naval forces.
22. David Brewster, "An Indian Sphere of Influence in the Indian Ocean?" *Security Challenges*, Vol. 6, No. 3 (2010), pp. 1–20, www.jstor.org/stable/26459796. (Accessed 17 June 2019).

6

Air Power for SAGAR

The IAF has had a remarkable and memorable journey since its formation on 08 October 1932. It has since been operating in an environment that is both challenging and demanding and consequently it is now a seasoned and dynamic air force.[1] To further justify this point, it is appropriate to go through the **global experience** (particularly in the IOR) of the IAF over 88 years of its illustrious existence and appreciate the journey it has undertaken so as to have an assessment of its capability to augment national vision about the Indian Ocean.

IAF in Congo

Much before the IAF was involved in the conflict against the sub-continental rivals in the 60s, its first taste of action came in faraway Africa, as part of a United Nations Force in Congo.[2] Though Democratic Republic of the Congo (DRC) is not part of the IOR, it brings out India's will and capability to help the friendly nations in need. The first time when the IAF helicopters were led in the volatile situation of Congo was in 1961. The IAF unit in Goma/Bunia was called the Indian Aviation Contingent (IAC-I).[3] During its deployment, IAC-I had achieved an elderly status among the various contingents deployed in Goma, being the oldest and most battle hardy. The environment for operating in such a complex scenario called for a range of innovative operational Standard Operating Procedures and survival strategies. In fact, majority of validation of helicopter combat tactics against ground threats were achieved through the lessons learnt in this foreign theatre. The other unit had Canberra bomber aircraft with aircrew and ground support personnel. In August 1961, the Indian Brigade Group had launched 'Operation Rumpunch' against the foreign mercenaries creating law and order problems in Katanga province. The Indian pilots used many ingenious methods to lead themselves to the targets[4] and made the operation successful. The IAF later took part in UN missions in Somalia and Sudan also.

Chetak Detachment to Batticoloa (Sri Lanka)

The IAF sent a cyclone relief detachment to Sri Lanka in November-December 1978 for about 17 days. Two Alouette helicopters were sent in An-12 transport aircraft and were reassembled and air tested the next day. The area affected by the cyclone where the crew conducted food parcel drops, stretched roughly from Kurunegala to Dambulla, Minneriya, Polonnaruwa, then the coastal area around Batticoloa, down to Amparai. They were cautioned not to fly close to Anuradhapura or Trincomalee (China Bay) since there was a danger of being fired on by small arms.[5] They flew many missions despite various challenges and returned once relief task was over.

The Sri Lankan Interlude 1987-90 (Operation Pawan)

The Indian involvement in Sri Lanka saw the largest air effort by the IAF in air maintenance and counter insurgency support amounting to 70,000 sorties by transport, helicopter and fighter aircraft without a single loss to hostile fire.[6] With the deteriorating situation in Sri Lanka's Jaffna Peninsula, the Indian government launched a massive effort to re-establish peace and normalcy through a two-pronged operation, flying in huge quantities of humanitarian assistance and a massive Peace Keeping Force. Here, the IAF played a major role of which Operation Poomalai is the best example. The supplies dropped amounted to not more than 23 tonnes and were not substantial at all. But it sent a message to the Sri Lankans that India would not be a bystander to the on-going civil war.[7]

Maldives Coup 1988 (Operation Cactus)

On the night of 03 November 1988, the IAF mounted special operation to airlift a parachute battalion group from Agra and fly it non-stop over 2000 km, out beyond the South-Western coast of India to the remote Indian Ocean archipelago of the Maldives. In response to Maldivian Government's appeal for military help against a mercenary invasion, the IL-76s of No. 44 Squadron landed at runway of Hulhule island at 0030 hours, the Indian paratroopers secured the airfield and Government rule at Male was restored within hours. Build-up of forces continued the next day, with IL-76s, An-12s and An-32s flown to the Maldives from Trivandrum, while the IAF Mirage 2000s made low level passes over the scattered islands in a show of force. The most immediate reality that emerged from this brief and bloodless action was the swift and effective Indian military response, made possible by the **IAF's strategic airlift capability**.[8]

IAF in Somalia

The IAF took part in peace keeping duties in Somalia from 01 October 1993 to 21 December 1994 as part of the Indian contingent supporting UN operations. The IAF helicopters were utilised for the following tasks:[9]

- Road opening and convoy escorting.
- Aerial Reconnaissance.

- Casualty evacuation.
- Communication.

Where few participating countries used force to achieve secured environment but failed, the Indian contingent used humanitarian approach. They took various projects like vocational training, vet training, food for work, adopting an orphanage, organising cultural functions and short events with local teams. This had an effect on unemployed population and they channelled their energies from banditry to useful work. This shows how Indian forces have the unique ability to stabilise situations of unrest in a foreign land without using hard power.[10]

IAF Contingent in Sudan

India continues to be one of the largest contributors to UN Peace Keeping Operations. The excellent performance of the IAF contingent in peacekeeping operations in Sudan has led to the UN asking for a rotation of the contingent. Modelled on the lines of a regular air force base, the IAF contingent's duty involved providing flying operations, administration, logistics, and technical and medical services support. The IAF contingent which went in year 2010 had six Mi-17 utility helicopters for troop insertion / extraction, casualty evacuation, logistic supply, search and rescue, reconnaissance and observation. The IAF contingent also maintained an aero medical evacuation team at Sudan.[11] An operation of the IAF (Operation Sankat Mochan) in view of 2016 Juba Clashes to evacuate more than 143 Indian citizens and other foreign nationals from South Sudan during the South Sudanese civil war was much appreciated.[12]

Role of the IAF in the Tsunami Relief (2004)

Operation Sea Wave

India launched Operation Sea Wave for relief, rescue and evacuation in the vast area affected by the deadliest tsunami of the century. On receipt of the first information about the disaster, at 0815 h on 26 December 2004, Air Headquarter swung into action and set its machinery in motion. A Disaster Relief Cell was set up at Air Headquarter Operations Room, on a round the clock basis.[13]

Operation Rainbow: Air Operations for Sri Lanka

At the request of the Sri Lankan Government, six medium lift helicopters were sent to Sri Lanka to undertake relief Operations. Of these, three helicopters were positioned on 27 December 2004 and three on 28 December 2004. The helicopters undertook Casualty Evacuation (CasEvac), distribution of relief supplies, deployment of medical teams and air dropping of food while operating from Katunayke and Minneriya bases. Total air effort including ferry of aircraft, was 445 missions, involving 316 hours of flying. The tonnage and passengers lifted were 329 tonnes and 882 respectively. The helicopters returned on 22 January 2005.[14]

Operation Castor: Air Operations for Maldives

Two para-drop modified, long range AVROs were tasked to proceed to Maldives on 28 December 2004 at the request of the Maldivian Government for assistance. These aircraft undertook inter-island operations within Maldives carrying out landings on various short field runways. Their task was to undertake CasEvac, air landing of food, water and other supplies and deployment of medical teams. Total air effort involved was 155 sorties involving 198 hours of flying. During this period 170 tonnes of stores and equipment and 885 passengers were air lifted.[15]

Operation Neer: Male (2014)

The city of Male (capital of Maldives) on 04 September 2014, suffered a major drinking water crisis with breakdown of its main RO plant. The city had to be sustained with over 100 tonnes of drinking water per day till the plant was re-operationalised. Based on a request by the Government of Maldives, the IAF responded with alacrity and deployed three C-17 and three IL-76 aircraft to airlift packaged water from Delhi to Arakkonam, and thereon to Male. The IAF airlifted 374 tonnes of drinking water to Male between 05 September and 07 September 2014.[16]

Operation Rahat (Yemen Evacuation 2015)

Yemen was pushed into a crisis situation due to developing civil war situation. Coalition Arab forces had commenced air strikes in Yemen in the last week of March 2015. The Government of India assessed a requirement of immediate evacuation of over 4000 Indian nationals based at various locations in Yemen. A combined MEA,

Fig. 6.1: Indian Citizens inside a C-17 aircraft of the IAF during Operation Rahat

Source: https://i.ndtvimg.com/i/2015-04/yemen-rescue-plane-650_650x400_71428465332.jpg

IAF, Indian Navy and Air India evacuation plan was put into place for evacuation of Indian nationals. While an Indian naval ship evacuated nationals from Yemen's port cities to Djibouti and Air India ferried back nationals from Sana to Djibouti, the IAF deployed three C-17 aircraft to ferry back Indian nationals from Djibouti to Kochi and Mumbai. As many as 11 evacuation trips were undertaken by the IAF aircraft that facilitated safe evacuation of 2096 Indian nationals.[17]

Operation Insaniyat (2017)

In response to the humanitarian crisis faced on account of the large influx of Rohingya refugees into Bangladesh, Government of India extended assistance to Bangladesh. The relief material consisted of items required urgently by the affected people, namely rice, pulses, sugar, salt, cooking oil, tea, ready to eat noodles, biscuits, mosquito nets etc. The relief material was delivered by the IAF planes.[18]

Other than these operations and missions carried out beyond Indian borders, the IAF has carried out countless relief and rescue operations within the country. Covering them is beyond the scope of this chapter; however, it is necessary to mention that the IAF is currently one of the most experienced forces to act promptly and carry out all kinds of operations due to ever persistent demanding circumstances like natural disasters, unconventional security threats and limited intensity conflicts imposed by hostile neighbourhood.

Why Air Power for Oceans?

Before narrowing down to what the IAF can do, it would be appropriate to understand what air power can do in the maritime domain (in IOR specifically). In just a century, the air power has evolved from flight of few seconds to flights covering the entire world without landing. Air power has become preferred military choice for rapid responses to enemy activity, the primary deterrent to aggression by adversaries, and a key enabler to military missions on both land and sea.[19] These military missions can be to project soft as well as hard power. In today's dynamic and challenging maritime environment, it is essential to have a clear understanding of the characteristics and limitations of air power vis-a-vis naval power. Then one can analyse the potential of air power to enable its optimum exploitation (in conjunction with naval power) for furthering national objectives. Even the doctrine of the IAF mentions that unless the unique attributes and limitations of air power are understood, its sub-optimal utilisation would continue repeating many instances of history, as air power is the youngest form of military power.[20]

Characteristics of Air Power

With the use of the third dimension which is the vertical dimension, air power possesses certain characteristics which are unique because of the factors embedded in the medium of transit like surprise, large coverage of view, concealment and obviously manoeuvre in elevation etc. These elements define certain unique attributes of the air power, like **flexibility** and **versatility** allow use of the same platform for multiple role and in ways suited to the situation anywhere across the spectrum. Speed and range gives **mobility** which makes air power the first element to be deployed in a mission. It was **responsiveness** which made the first aircraft get airborne within five hours of hitting Tsunami in 2004 and within 16 hours of receiving the SOS call from Maldives in 1988. **Reach** is another inherent characteristic of air power because of use of third dimension. **Offensive action, shock effect and concentration** are other attributes which make air power germane to all kinds of situations ranging from peace keeping to nation building while deterring challenges to national interests.[21]

While these are unique attributes of air power, few limitations which are relative rather than absolute should be looked at to have a holistic understanding of air power. It is important to understand that while the strengths of air power can augment other elements of national security or measures of nation building, the weaknesses of air power can be overcome with judicious use of assets and appropriate division of responsibilities. The limitations of air power are sustainability, base dependency, sensitivity to technology, vulnerability, impermanency, weather and political constraints. In the context of the IOR, most of these limitations can be compensated by good planning, better coordination with the Indian Navy and the Army and enhanced international defence cooperation. This can be better understood by looking at characteristics of navies which make them suitable for maritime diplomacy. The attribute of versatility, mobility, reach and offensive posture are same as air power's attributes but uniqueness lies in characteristics of independence of movement, endurance and graduated response as well as connectivity of the seas where in a naval fleet does not require any base support, has ability to hold the position for long durations, respond to situations with desired pace and scale while being in seas and have natural access to and from all coastal areas.[22] These characteristics make the Navy most apt service for some particular situations.

Similarly, the unique attribute of swift response through the medium of air makes the air force best suited for many situations. Uniqueness of these two services can be mutually complimentary for achieving common aims. For example, in case of natural disasters in any nearby IOR nation, rescue teams of Indian Army or National Disaster Relief Force (NDRF) can be dispatched in a much shorter time by IAF's C-17 aircraft along with relief material and medical team. These aircraft can also evacuate people who are in urgent need of medical help followed by quick re-supplies. The Indian naval ships can augment the HADR assistance subsequently and provide further help. However, the arrival of help in shortest time by the IAF aircraft is a definite way of nation building and projection of soft power, as it happened in the Tsunami of year 2004. If there is a threat of piracy, hijacking of cargo ships or terror attack through seas, while the naval craft approaches the scene, the effect of a fighter aircraft flying by can deter any eventuality. As it happened in Maldives during Operation Cactus in 1988 wherein only the news of the IAF aircraft landing at Male airport in pitch dark night within 24 hours made the rebels run and eventually surrender to Indian forces. Thus, the characteristics of speed, range, responsiveness and mobility in third dimension makes the air power an indispensable tool for promoting soft power as well as safeguarding national interests of India and other IOR countries.

If the attributes (strengths) and weaknesses of both the services are put together, they can be complimentary. The strengths of the air power can overcome weaknesses of naval power and vice versa. It is to be understood in backdrop of optimum utilisation of capabilities and in turn, of assets. For example, the CBG of any country is considered a strong representation of that country's might in the oceans. Over the years the CBGs have become large flotillas and the available fighter air effort are fully extended in providing air defence to the ships of the flotilla. If one studies the US naval carrier employment after the 1980s, one would see that these forces became increasingly dependent upon long range land based air forces for a whole gamut of support including aerial refuelling, reconnaissance, air borne warning and control, Electronic Warfare (EW) and Suppression of Enemy Air Defences (SEAD). New developments in long-range precision guided munitions and target-locating systems have added much needed potency to land based aircraft. Availability of aerial refuelling capabilities have largely extended the operational envelope of these aircraft, thereby making land based aircraft increasingly effective in the maritime warfare role and compelling CBGs to operate further away from shore.[23]

IAF and the Maritime Air Operations

It is important to know at this point that as per basic doctrine of the IAF, the overall responsibility for the air defence of the country has been vested in the IAF, along with other areas including the island territories and Off-shore Development Areas (ODAs). The defence of the airspace at sea within the limits of Air Defence Identification Zones (ADIZs) except the AD of naval assets at sea also rests with the IAF. The AD of naval assets at sea, within the ADIZ or outside is a naval responsibility except when specific assistance is sought from the IAF.[24]

For Maritime Air Operations (MAO), the IAF is employed in close cooperation with Indian naval forces including naval aviation assets. Control of the IAF aircraft is exercised by HQ, MAO in coordination with the Indian Navy. The two defined maritime air roles for the IAF as per the basic doctrine are Anti-Shipping strikes and Maritime strikes. While both operations need long range fighter aircraft, AWACS, FRAs and ISR platforms of the IAF along with naval assets for coordination, the targets for the IAF are enemy's naval assets in the vicinity of own naval forces in the former and enemy's assets that are not in contact but pose indirect or long term threat like Maritime Patrol Aircraft, harbours etc. in the latter case.[25]

Another important but less mentioned role of the IAF is to ensure certain degree of control of air for maritime operations being conducted within extended range of land based air assets. The practice of providing suitable degree of control of air[26] for naval operation by shore based long range aircraft is common doctrine for many countries. The active involvement of the IAF in projection of India's stature in defending and securing the IOR is thus necessary.

In the Indian scenario, the air wing of the Indian Navy has many assets and their deployment is mostly independent. They are not deployed in the manner (operationally or strategically), in which the IAF deploys its assets. This difference is logical because the roles and area of operations are different. But there are always options of exploiting the possibilities of enhancing interaction of air power proponents of India in the IOR. The aircraft carrier can be seen to be useful in a tactical scenario to create deterrence in a particular (battle) zone with fighter aircraft (MiG-29K) on board for protection of CBG from aerial threats but with increased range and speed of modern anti-ship weapons, the range of these aircraft has relatively reduced. World is now witnessing the increasing proliferation of supersonic anti-carrier cruise missiles with ranges upward of 500 km which may not be able to sink a carrier but can put it out of operations for a certain period.[27]

For in-flight refuelling (to increase the range of aircraft), the Navy practices buddy refuelling drills i.e. one aircraft carries refueller pod and other aircraft practice taking fuel from that aircraft. While sailing in the open seas and carrying out operations from carrier, one fighter aircraft takes off in the last from the deck with fuel pods to let others land safely on the carrier in case of missed landings or other emergencies in which extra fuel may be required. But this appears to be a make-shift arrangement. The FRAs of the IAF are made available on demand for practicing in-flight refuelling of naval aircraft but since the FRAs are limited, this frequency is less. The carrier operations should be done with FRA in the air and carrier based aircraft should benefit from this. Thus more number of FRAs should be procured. IAF fighters also can be used to sanitise the threat which is beyond the range of shipborne aircraft (their range is phenomenal if they take off from a coastal base or from islands, refuel in air thus increasing range further). The IAF aircraft can also simulate as adversary carrier strike force and can add to training value of air defence of the CBG. Joint exercises between the IAF and the Indian Navy should increase to maximise the exposure on both sides and optimise the effectiveness of air power in IOR. On the other hand, given the compelling need of optimum utilisation of defence budget, it is unwise to burden only the Indian Navy for guarding the Indian Ocean. The Navy has one carrier as of now and another is under construction. The Navy has understandable need of more aircraft carriers but then it will need more fighter aircraft on deck to protect the CBG. If there is requirement for more fighter aircraft for the Navy, it shall be looked at with carefully calculated requirement of the number of aircraft carriers India needs in future.[28] The defence budget limitations and future trends also need to be catered for.[29]

At this point it is important to have a look at the present capability of the IAF in order to understand what it can do to secure the IOR and what can be done to make it even more suitable source of nation building. For that it is necessary to know the present holding of flying assets (manned) of the IAF which is shown in Table 1.[30]

From Table 1 it becomes clear that the IAF has a mixed inventory of various types of aircraft in each category. It thus has the ability to perform all kinds of tasks required not only for sustaining IAF operations but also tasks like logistics support to Army or evacuations, for example. It is said that an air power whose doctrine, training and equipment heavily favour fire power over logistical mobility, intelligence gathering, medical evacuation, and so forth unintentionally is likely to impose damaging opportunity costs on its political owners.[31]

The IAF has, in this way, gained a unique ability to operate in all scenarios depending on the demand of the situation. Though all types of aircraft have their special role in projecting India's positive image in the IOR, capabilities of few aircraft which can prove to be crucial in times of need are mentioned after Table 1.

Table 6.1

S No	Category	Name	In service	Remarks
1	Fighter/Attack	SU_30MKI	242	8 ordered
2		Mirage 2000	45	
3		MiG-21	132	To be phased out
4		Rafale	1	35 ordered
5		MiG-29	66	
6		Jaguar	130	
7		LCA Mk 1	16	99 ordered
8	Transport	IL-76	17	
9		An-32	104	
10		Avro	57	To be phased out
11		Dornier	50	
12		C-130 J	11	
13		C-17 Globemaster III	11	
14		Embraer	3	
15		Boeing 737	2	
16	Tanker/Refueller	IL-78	6	
17	Trainer	HJT -16 Kiran	78	
18		Hawk Mk.132 AJT	104	
19		PC-7 MkII	75	
20	AEWC	Embraer ERJ-145	2	
21	AWACS	A-50 E	3	
22	Helicopter	Mi-8	80	
23		Mi-17/ V5	223	
24		Mi-26	1	
25		Mi-35	15	
26		Cheetah	24	
27		Chetak	48	
28		Dhruv	91	
29		AH-64 E	8	
30		CH-47F	6	

C-17 Globemaster

It is a high wing, four engine aircraft with wide body and rear loading ramp. This aircraft has already participated in many operations like Operation Neer and Rahat. With its fast loading capability and

Fig. 6.2: C-17 Globemaster III Strategic Airlifter of the IAF

Source: www.defpost.com

massive lift capacity, it can be used in any kind of situation. It can deliver relief material or rescue teams in times of disasters and evacuate patients/casualties in large number from anywhere in the IOR. If needed, it can do insertion and extraction of personnel while operating from unpaved surfaces also. Its maximum load capacity is 75 tonnes and range is 4500 km while with 45 tonnes load, range extends to 7400 km without refuelling (this aircraft has aerial refuelling capability). It can carry 102 paratroopers or 188 seated passengers or 102 ambulatory patients.[32]

C-130 J Super Hercules

It is a tactical transport aircraft made for special roles like Special Forces insertion, landing at small and unprepared airfield, faster loading and offloading and turnaround time. It has a maximum payload capacity of 24 tonnes and can carry 128 combat troops or 92 paratroopers/97 medical litters.[33] The range with 18 tonnes payload is approximately 5250 km.[34]

Fig. 6.3: C-130J Super Hercules of the IAF

Source: www.defpost.com

Airborne Warning And Control System (AWACS)

This platform has the most advanced sensor systems and new age technologies which make it an airborne Command, Control, Communications, Computers, Intelligence, Surveillance and Reconnaissance (C^4ISR) centre and thus a valuable force multiplier. Capable of detecting airborne and maritime targets as well as providing data link communication to tactical elements, the AWACS has unique capability of enhanced detection by virtue of height of the sensors which increases low level detection range of the radar and thereby produces early warning. By capability of responding quickly and remaining airborne for long time, it has flexibility and versatility and is less vulnerable because of mobility. If the in-flight refuelling practice for AWACS is put in place (for which the IAF needs more assets), it can be deployed over extended and inaccessible terrains such as seas, deserts and other inhospitable terrain. Other than role of controlling large packages of various types of aircraft and gathering air intelligence along with ensuring strong air defence, AWACS is also very helpful in the role of HADR. Post natural disasters, it can play a vital role in surveillance, aircraft control and airspace management to fill the void created by inability of ground based systems to move in quickly and operate in difficult terrain

Fig. 6.4: An IAF AWACS (on an IL-76 Aircraft platform) escorted by MiG 29s

Source: www.asiapacificdefensejournal.com

and poor infrastructure. It can also help establish localised command and control mechanisms until ground agencies are able to build situational awareness. Another important role of AWACS is to provide search and rescue (SAR) support through direct communication with non-traditional joint players such as coast guard, local enforcement and rescue teams of any country.[35]

Flight Refuelling Aircraft (FRA)

Since 2003, the IAF is operating IL-78 tanker aircraft which have enabled the fighter aircraft of the IAF stay airborne longer for operational requirements as well as reach many countries across all oceans. These aircraft considerably reduce the lead-time and cost of operations while increasing the number of options available for operations. This capability (though with limited number of FRAs) is of particular relevance for India in order to assist maximum IOR countries in time of need as it gives advantages of increased range, increased endurance, improved response time and more flexibility.[36] With increased number of FRAs (with capability of refuelling transport aircraft and AWACS also), the IAF can plan short halt missions at far flung IOR nations as well, with least logistical support. FRAs can also greatly boost the capability of the Navy to operate their P-8I and MiG-29K aircraft for longer durations and thus at increased ranges.

Fig. 6.5: FRA (on an IL-78 Aircraft) of the IAF

Source: https://www.aircraftcompare.com/blog/how-do-airplanes-refuel-in-the-air/

These are the force multipliers of the IAF which make it capable of providing an assurance for not only India's security from all kinds of threats but for all the countries of the IOR who believe in the concept of mutual respect and help. Chief of the Air Staff recently brought out importance of the IAF in the IOR in terms of load carrying capability for regional support. He said that the IAF has enhanced its airlift capability through C-17 and C-130 aircraft and it can take 300 tonnes up to a distance of 6000 km in one wave and can repeat it practically every six hours. And if shorter ranges in neighbourhood are considered, it can lift 600 tonnes up to 3000 km in one wave. Also in terms of helicopter support, the IAF can provide any number of them. Thus in terms of HADR and providing regional support, capability of the IAF is huge.[37] Along with heavy lift aircraft, Indian fighter jets can also reach to support any nation in the IOR, under gap free surveillance of AWACS and with in-flight refuelling by FRAs, if and when required (be it for bilateral exercise or in case of an emergency).

Other than the capable fleets of aircraft, the technologically advanced IAF has other assets and force enablers that give it the edge and niche for being the most advanced air force in the region. That includes radar systems (for all kinds of terrains and detection ranges), Surface to Air Missiles (SAM) to tackle all kinds of airborne threats and new age Remotely Piloted Aircraft (RPA) for the role of Reconnaissance, Surveillance and Target Acquisition (RSTA) along with maritime operations, EW, Signal Intelligence (SIGINT), meteorology missions, battle damage assessment, lasing the target systems for precision attacks etc. The IAF also has Integrated Air Command and Control System (IACCS) as part of its net centricity, where in air picture of all military and civil sensors are available at one place to reduce reaction time and enable quick response to impending challenge by reducing sensor to shooter time. These capabilities of the IAF give India the potential to provide the IOR countries all kind of assistance, support and cooperation to overcome various challenges which can pose a threat to the growth and security of the region.

Reach of the IAF in the IOR

India has ability to launch its aircraft and assets from airfields located at various locations. Important airfields with their distance from key locations in the IOR are given in Table 6.2.

TABLE 6.2

Sl No	*Area of Influence*	*Chennai (km)*	*Coimbatore (km)*	*Lakshadweep (km)*	*Port Blair (km)*	*Jorhat (km)*	*Imphal (km)*
1	Hanoi	2854			1762	1343	1290
2	Bangkok	2187			897	1584	1406
3	Kuala Lumpur	2599			1386	2734	2539
4	Yangon	1753			688	1125	920
5	Laos	2451			1291	1301	1175
6	Phnom Penh	2682			1354	1945	1860
7	Singapore	2906			1700		
8	Jakarta	3622			2954		
9	Manila	4397			3765	3088	3054
10	Maldives	1236	900	714	2250		
11	Diego Garcia	2420	2090	1980			
12	Seychelles	3374	2954	2538			
13	Comoros	4917	4488	4068			
14	Madagascar	5046	4626	4269			
15	Mauritius		4054	3781			

Note: All the distances are less than 6000 km as mentioned by the CAS in a recent public interaction.[38]

Additionally the IAF has fully operationalised Air Force Station, Thanjavur in Tamil Nadu and one squadron of Su-30 aircraft has recently been based there.[39] This base is 350 km south of Chennai which will further expand reach in the IOR and help protect India's island territories and SLOCs. However, the fact should not be ignored that the best choice to operate the Air Power assets (in situations demanding constant presence) are island territories of India (especially Andaman & Nicobar Islands) for projecting its presence over IOR's important SLOCs. Hence it is necessary to develop these islands which are strategically very important, for staging and sustaining operations by the IAF. Figure 6.6 shows the approximate range of a medium size, tactical lift aircraft without refuelling (the least range of 3000 km is depicted whereas C-17 and C-130 have higher ranges). Three circles which are numbered from 1 to 3 show reach of aircraft getting airborne from Thanjavur, Car Nicobar and Minicoy respectively.

Strategic Potential of Islands

It is evident from the previous table that the distance to almost all IOR nations reduces significantly if measured from Port Blair (Andaman and Nicobar Islands) or Agatti (Lakshadweep). It will be further reduced if respective southernmost islands i.e. Great Nicobar and Minicoy have adequate infrastructure to base and operate aircraft of the IAF. Indian naval base Baaz is operational at the Great Nicobar's Campbell Bay as India's southernmost and easternmost naval air station but with a very short runway with length of only 3500 feet.[40]

Andaman & Nicobar Islands and Lakshadweep islands are like biggest stationary aircraft carriers as well as natural barriers for India. These islands can provide many military options to keep the busiest sea lanes under surveillance and to have projection in the Bay of Bengal.[41] It allows India to keep a vigilant eye on the strategically important Malacca Strait, through which 80 per cent of China's crude and oil imports from the Middle East and Africa passes, along with fuel imports for Japan and South Korea.

Presently, the IAF with its limited assets is not represented adequately in the Andaman and Nicobar Command (ANC) which is the first and only tri-service theatre command of the Indian Armed Forces, based at Port Blair created in 2001 to safeguard India's strategic interests in Southeast Asia and the Strait of Malacca by

Fig. 6.6: Reach of a Medium Size Tactical Lift Aircraft in IOR Airborne from India

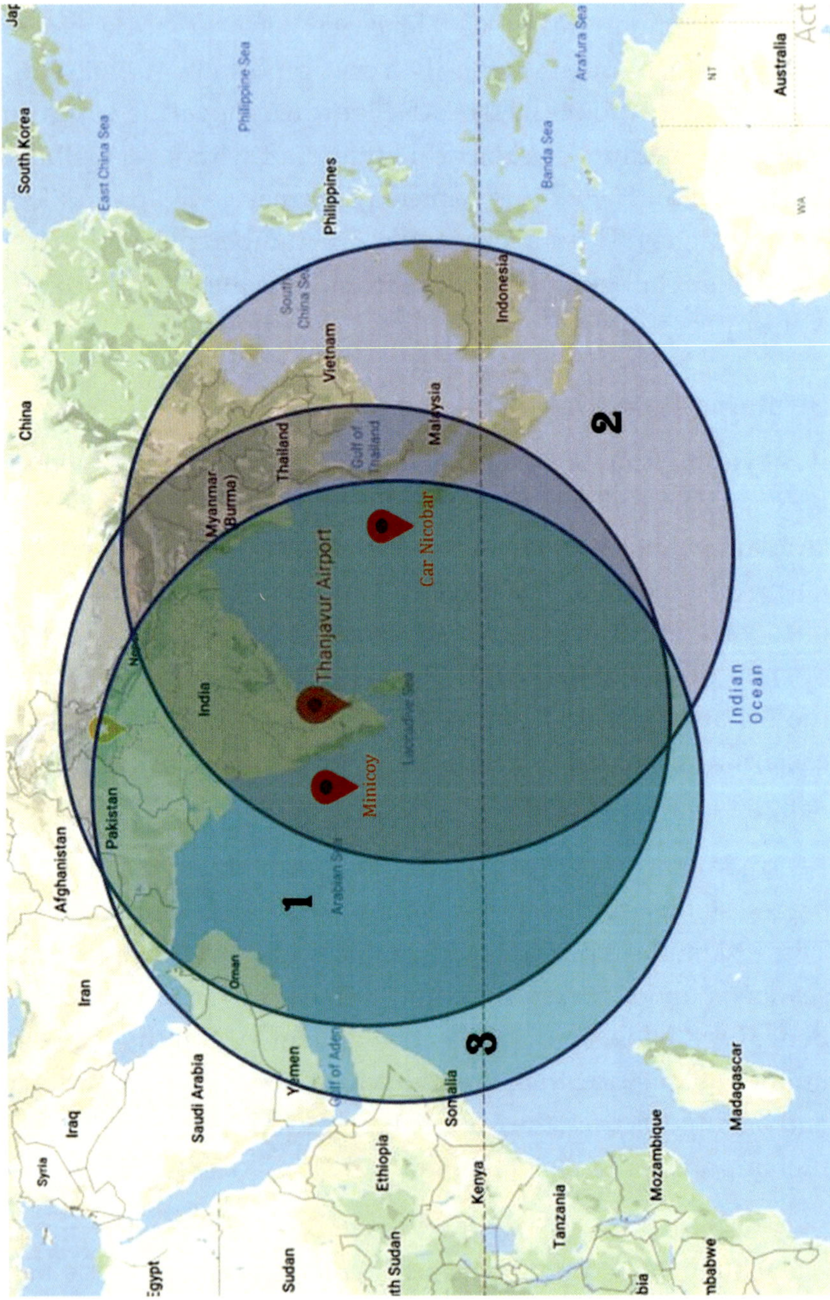

increasing rapid deployment of military assets in the region. It has deployed five Mi-17 V5 helicopters at an airfield in Car Nicobar which has a runway with rough surface, narrow taxi track and only one or two radar around. The logistics support is given by mainland based transport aircraft on regular basis.[42] These assets are useful not only for military purpose but also for helping islanders in medical emergency, urgent evacuation or supplying ration in case of natural disasters. Given the fact that these sentinel islands need to be strengthened militarily, there is a need to augment present air power support system.

On the western side i.e. Lakshadweep, the air connectivity is less developed as compared to ANC. As of now, only Agatti Island has one small airstrip with very basic support to operate small aircraft with turbo-prop engines. Rest of the islands are accessible by helicopter but that connectivity is also limited due to lack of infrastructure. There is a proposal of developing a dual use air base at Minicoy Island which is the southernmost island only 70 miles away from northernmost island of Maldives. It is the biggest island in Lakshadweep and also one of the most visited islands by tourists from all over the world every year. The survey has been completed long back but the ground work is yet to start. This island has lot of potential and the airport will improve tourism, connectivity and employment. With the developed infrastructure, operational transit of the IAF aircraft also can be managed. The possibility of an air base with longer runway for operating heavier military planes from the IAF fleet, such as C-17 and C-130J should be looked into.

These projects need to be prioritised as these islands will prove to be India's Eastern and Western frontiers as well as launch pads to serve the purpose of the regional security provider. China has been using its islands for military purposes for a long time. The front tier of islands reclaimed from sea and development of airfields to stage H-6 strategic bomber aircraft[43] along with strong air defence system by China has set an example of importance of land based air power in its doctrine of Anti-Access/Area Denial (A2/AD).

Exploitation of the full potential of existing military set up under ANC suffers from serious deficiencies like overall prioritisation by the MoD as well as the asset allocation by respective service headquarters. Lack of jointness between service headquarters, lack of infrastructure development and lack of bare minimum force structure[44] has been the main hindrance for 19 years since its establishment. The attitude for capitalising the worth of this joint structure under established integrated command is missing. The key role of newly appointed Chief of Defence Staff (CDS) is to bring about jointness in operations, logistics, training, support services, communications, repairs and maintenance.[45] All these areas in the ANC need attention and it is expected that the office of the newly appointed CDS shall do the needful in near future. Given the stakes of India in the IOR, a revamp of the structure for advanced military set up at both island territories needs consideration, and the name from ANC could be changed to the IOR Command with the IAF having complete responsibility of providing air cover by employing all air assets in an optimum way using jointness. The ADIZ around the island territories should be extended after due consultation with other affected IOR countries so as to have extended AD cover over India's AOR.

IAF for the IOR

From the examples of the IAF's operations abroad in past, unique characteristics of air power and present capabilities, it is clear that it is a seasoned air force with a strong doctrine to complete any task given to it with grit and commitment. The tasks that the IAF can undertake to promote India's aerial diplomacy and should be encouraged to do so in the IOR are:

- **Securing Sea Lanes of Communication.** The protection of shipping lanes from attack by hostile states or irregular threats can be provided by the IAF depending on the requirement. The aerial patrolling of SLOC is normally carried out by maritime patrol aircraft but this effort can be supplemented by Su-30 MKI aircraft that have an extended range and loitering capability[46] with AWACS and FRA in air if the situation so demands.
- **Non Combatant Evacuation Operations.** The evacuation of Indians or citizens of friendly nations from a foreign country when they are endangered by war or civil unrest can be carried out in minimum reaction time possible. This aerial diplomacy has been demonstrated on many occasions in the past by the IAF. Given the current and ever increasing global spread of the Indian diaspora (the latest census according to MEA is 32,100,340)[47], a situation requiring evacuation could arise any time. The IAF with other services, due to its faster response time, is always ready to execute this task as and when situation arises.
- **Humanitarian Assistance and Disaster Relief.** The speedy deployment of the IAF assets to mitigate the impact of natural disasters has been an important reason in reducing the casualties. The IAF has set many examples as an efficient first responder within and outside the country. The heavy lift capability and responsiveness of the IAF are very vital to not only Indian civil HADR system but also to many neighbouring IOR countries who have witnessed the aerial help reaching in time of crisis. In these situations, the IAF is primary source of nation building.

In remote Indian areas, Mobile Medical Teams (MMT) are sometimes carried by the IAF aircraft in times of contingencies. The IAF formulated its own Critical Care Air Transport Team (CCATT) in 2007, using an indigenously designed Patient Transfer Unit, which could be retrofitted in either fixed wing or rotary wing aircraft. As many as 17 transport aircraft of the IAF can be converted into Air Ambulances.[48] This assistance in terms of training and knowledge transfer can be extended to other IOR countries also as and when needed. Search and Rescue (SAR) missions for saving lives at open sea also form part of humanitarian assistance and case of Commander Abhilash Tomy should be looked into for exploiting possibilities.

Case of Abhilash Tomy: Commander Abhilash Tomy of the Indian Navy was rescued from the southern Indian Ocean (around 1,900 miles southwest of Perth, Australia) on 24 September 2018 after being stranded for three days. Abhilash, who was taking part in the Golden Globe Race (a circumnavigation tournament), was caught in a storm on 21 September 2018 that wrecked his boat and rendered him immobile due to severe back injury. In his distress messages, he had indicated that he was confined to his bunk bed inside the boat and was finding it difficult to eat or drink anything. An international mission was launched to rescue the Indian sailor and on 23 September, Indian Navy's LRMP aircraft P-8I visually contacted Abhilash Tomy's boat.[49] Finally, a French patrol boat picked him up and took him to nearest French territory – Amsterdam Island which had a medical set up.

For such contingencies, there is an option of joint operation which was not tried then. Indian Navy's Marine Commandos (MARCOs) are well trained for para jumps from aircraft into the sea. An IAF aircraft could have air dropped team of MARCOs with inflatable rubberised boat, adequate ration and necessary first aid items near the location, once it was known. The sight of Indian forces itself would have given Commander Tomy a sigh of relief and the first aid along with food would have made his 3-day long ordeal much shorter.

- **Peacekeeping Missions:** Participation by the IAF in the UN mandated peace keeping missions have furthered India's reputation as a responsible regional player. As brought out previously, tasks carried out by the IAF contingents in Congo, Sudan, Sierra Leone etc. have left unmatched impressions. Presently, the UN cell in the Army Headquarters is the central point of contact and detailing for all UN missions. The IAF has active representation in few missions of UN but the engagement has scope of improvement if three services have detailing done by the HQ, IDS which will ensure active involvement of all services as per their resource availability. The IAF is capable to undertake peace keeping missions anywhere in the IOR under the flag of the UN depending on the scale of effort.
- **Protecting Democracies in the IOR:** There have been instances wherein speedy interventions by the IAF, on specific requests from friendly countries to the Indian government for assistance to ward off risk of human conflict and threat to sovereignty, have averted danger to diplomacies in the region. The speed and responsiveness of the IAF facilitates the government to show its intent for keeping the region secure from challenges to democratic countries of the IOR.
- **Sub Conventional Operations:** The flexibility and precision of air power makes it eminently suited for sub conventional operations. The proliferation of terrorism by non-state actors has become a norm to target soft underbelly of democratic nations. The IAF can play a major role in bolstering political will by interdicting terrorist/insurgent feeder camps, training camps, supply nodes or operating bases. The IAF can perform all roles efficiently in such scenarios like surveillance and reconnaissance by all

available platforms, air mobility and logistics support, air defence cover, targeting precisely with minimum collateral damage and casualty evacuation if required. The IAF has proven its calibre in Balakot surgical strike in February 2019.

- **Out Of Area Contingency (OOAC) Operations:** As India's economy grows and its area of influence increases, the defence forces have to prepare for tackling any contingency that has the potential to adversely impact national objectives and occurs beyond geographical boundaries. In such scenario there will be need of speedy mobility, effective surveillance, precision engagement and support for re-supply of troops and equipment. Therefore, the IAF will form an integral part of most task forces with help of its diverse airlift platforms, FRAs, long range fighters and if possible, it will be required to operate from air bases of other friendly countries. Though the IAF is ready to actively participate in OOAC operations as part of a joint Indian task force, there is a need to validate these concepts through realistic exercises[50] and more defence Memorandums of Understanding (MoU) with friendly neighbours. With the announcement of charter of responsibilities for newly appointed CDS, more OOAC operations shall be taking place in near future and the IAF will play a crucial role in the domain of military diplomacy.
- **Capacity Building:** The IAF has gained enough experience and technological expertise to support other IOR nations in capacity building and capability enhancement. Many 'Made in India' assets which can enhance air power diplomacy can be exported to friendly countries and the IAF can be a consultancy and service provider by helping those countries through stages of supply, set up, training and independent operations. For example, the Indian Air

Force has helped Sri Lanka in setting up of AD radars when requested. Team of professionals was also sent to set up the systems and train the users operationally. Now when India has many good indigenous radars, it can extend this help to make its strategic neighbours secured from aerial threats and more aware of aerial activities in the IOR. AD networks, communication networks like IACCS and AFNET are such examples. Entering into AD data sharing agreement will enhance the aerial awareness of the region. Such efforts can not only boost the nation building factor but also bring other countries on same platform in terms of common understanding between air forces.

- **Joint Exercises:** As brought out earlier, the IAF has been engaging in friendly exercises with many countries but with IOR nations, the number of such engagements is less. The IAF version of PASSEX involving only few aircraft landing after a direct flight for short duration transit aimed at exchange of experience can boost the confidence level of the IOR nations. This will also assist in exchange of training and operational philosophy. These short duration, small scale exercises will not involve excessive planning or huge expenditures and will gradually create good-will between air forces of the IOR. A multilateral platform of IOR nations for joint aerial exercises based on different themes is another concept that can be looked into. The frequency of exercises and number of participating countries can be decided by a formal committee and area of operations shall be kept variable so that entire IOR is covered. These interactions will boost the security paradigm of the IOR significantly.
- **Air Shows:** The IAF can promote indigenous defence products at international air shows of the region. Advanced Light Helicopter (ALH), LCA and LCH are well proven

indigenous products and the IAF can boost military-industrial ties at premier air shows like LIMA (in Malaysia), African Aerospace and Defence (AAD) expo in South Africa, Australian international air show, Singapore air show and Dubai air show etc. The IAF participated in LIMA-2019 with two LCAs in March 2019 for the first time and the aerial display by LCA was well appreciated.[51] This kind of participation has scope of growth in terms of involvement of leadership to promote capability of the IAF, presence of aircraft and technical delegations to support export offers. This will enhance military to military ties as well as industrial ties in the region.

The IAF has conducted air displays by aerobatic teams in few countries. Surya-Kiran Aerobatics Team (SKAT) of the IAF has gained many laurels across the world by its fascinating air shows. Having a nine ac formation leaving tricolour smoke in the sky itself can be a friendship booster. Similarly, Sarang helicopter formation display team can leave good imprints in the minds of the people. Low level aerobatics by fighters at the end of such air shows will surely gather many praises for the IAF. A social, cultural or even a sports meet during such interactions will add to the quality of such events.

- **Air Domain Awareness:** Peace time cooperation such as coordinated shadowing or shepherding of a simulated rogue aircraft, sharing radar tracks or air pictures etc. can be done on regular basis. On lines of sharing the MDA through Information Fusion Centre, the air pictures also can be shared through multiple radars and sensors as well as satellites which can prevent mysterious loss of passenger aircraft like MH-370 from Malaysia and transport aircraft like An-32 of the IAF from Chennai. The air picture fusion centre will have civilian and military controllers who will keep sharing the required information with the aircraft operators as well as authorities to keep track of all aerial activities in the IOR. This can also be achieved by merging ADIZs or creating new ADIZs with well divided responsibilities.

- **Defence Space Agency:** The tri-services Defence Space Agency (DSA) is going to be headquartered in Bengaluru. This agency is likely to command all the space assets of the three services.[52] With the IAF as the lead service, India needs to enhance utilisation of space for the defence of national assets and shall consider establishing an Aerospace command to have better projection in the fourth dimension. India has also completed the Indian Regional Navigation Satellite System (IRNSS) constellation for providing accurate position information to users in India as well as the region extending up to 1500 km from its boundary, which is its primary service area. An extended service area lies between primary service area and area enclosed by the rectangle from Latitude 30 deg South to 50 deg North, Longitude 30 deg East to 130 deg East.[53] This constellation can be used to aid in navigation for missions on land, in sea and in air; and can very well serve as an invaluable component of network-centric warfare.[54] After complete activation of the ground segment, IRNSS based applications will be useful for terrestrial, aerial and marine navigation, disaster management, vehicle tracking and fleet management. Its integration with mobile phones will make it an indigenous replacement of the American system mostly in use called Global Positioning System (GPS).[55] As shown in the Fig. 6.7, the coverage of this system will extend in the entire IOR, hence this technological advantage can become another platform for the countries of the region to come together with India and benefit from accurate navigation information which is the foundation of many crucial services.

Fig. 6.7: Coverage Area of the Initial Seven-Satellite IRNSS System[56]

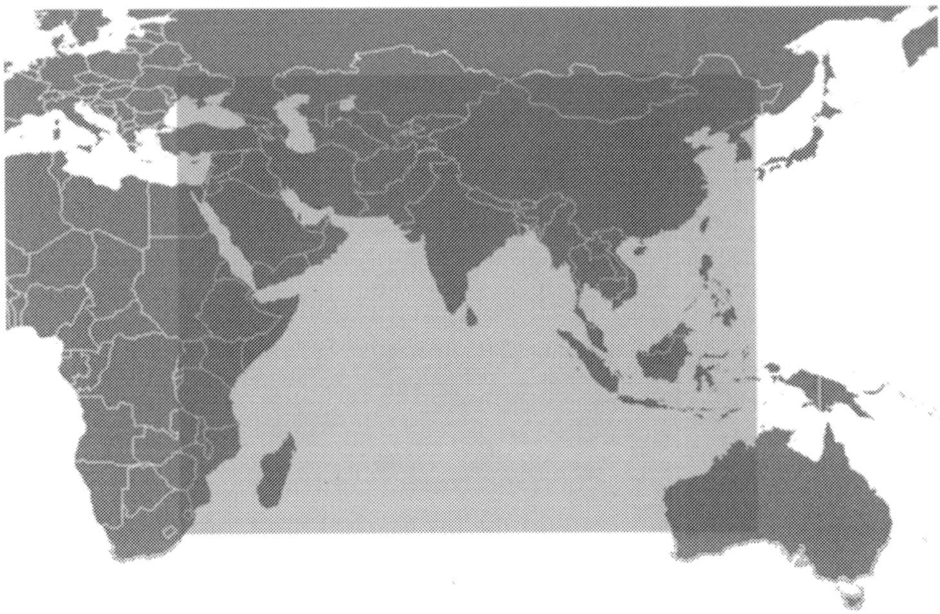

Source: Brilliantwiki2

Scope of Improvement

- To have a better control of air over mainland as well in the IOR, the IAF must invest more in force multipliers like AWACS, FRA, networking platforms and secured communication technology. Early warning aircraft will not only enhance India's air defence radius but will also play a key role in any expeditionary context. Similarly, refuelling aircraft will increase the range and weapon loads of strike aircraft thereby adding to India's air deterrent.[57] AWACS aircraft of the IAF has AAR capability but due to very limited numbers available, the operational risk is very high in using this capability which can keep AWACS airborne for longer duration. Strategic bombers, like nuclear submarines are also force multipliers which can be looked at for procurement as joint strategic assets.

- Planned investment in Intelligence, Surveillance and Reconnaissance platforms like RPAs and satellites and their integration for overall improved intelligence apparatus is necessary for enhancing India's role of regional security provider. Sharing of intelligence between the military and intelligence agencies remains suboptimal, and India currently lacks a robust C^4ISR system suitable for network-centric operations.[58]

- The Navy has furthered its role by close interaction with other navies (given the opportunities they get by sailing into international waters). They are more aware of the defence requirements of other countries and even other aspects of international relations as compared to the IAF. There is a need to generate such opportunities for the IAF to interact with other air forces/wings. India can help smaller countries develop their air wings after thorough diplomatic discussions on such topic. Countries like Madagascar, Comoros (which are now part of the IOR division in the MEA) and others which need training (in flying, meteorology etc.), equipment, and knowledge transfer on certain set ups like Modernisation of Airfield Infrastructure (MAFI) project, should be approached and engaged more actively. For example, if Mozambique needs help from the IAF to patrol their skies and provide assistance in aerial survey, India must capitalise on such opportunities. The IAF must also pitch and promote indigenous platforms like LCA, LCH, ALH etc. by means of demonstration while the government should make arrangements on diplomatic level for such defence deals.

- Selected officers of the IAF must be sent to various divisions of MEA on deputation (the vacancies are regularly published on website of MEA).[59] They should be selected on basis of their interest and knowledge in international affairs and their performance in staff courses (subject of area studies). Officers who are doing or have done some research specific to a zone/country should be encouraged. It will develop close liaison between organisations and ensure benefits on both sides by reducing time and effort in flow of information and necessary actions. The list of defence officers (as on 30 December 2020) [60] on deputation to various divisions of MEA is attached as appendix 2. Out of 17 officers, 14 are from the Army, three are from the Navy and **none from the IAF**. In the interview with the Joint Secretary of IOR division, it was found out that there is requirement of defence officers in such divisions handling external affairs with countries having active military diplomacy with India.[61]
- Air HQ has directorate of intelligence as the main directorate for collating intelligence as well as to manage foreign cooperation through a sub directorate which looks after visits and exchanges with armed forces of foreign countries. In order to give impetus to international engagement of the IAF, there should be one dedicated directorate for international cooperation with an air officer as in-charge. The existing sub directorate can be changed into a specific directorate of foreign cooperation with dedicated region wise officers to handle deputations, engagements and interactions based on air staff interactions or ministerial instructions. The present manpower can be augmented with additional personnel from relevant directorates having direct involvement in international

cooperation like directorate of training, AD, plans, procurement as well maintenance to enhance their relevant areas of foreign interaction.

After having seen the present capabilities of the IAF in maritime domain and its potential to further the Indian vision of security and growth for all in the IOR i.e. SAGAR, significance of the IAF and enhanced Air Power in the IOR must be seen in conjunction with the importance of a true blue water Indian Navy in the IOR. Because in the Indian context, it is unlikely that any future hostile maritime conflict would take place far from land-based air forces. It would therefore be inconceivable to undertake any naval action without ensuring the availability of favourable air situation to support the action. This is particularly true for air superiority operations near the coast of an adversary where an exclusively naval force may be unable to operate freely, or where an enemy's anticipated air defences may pose grave dangers to conventional naval attackers.[62]

Given the geo-strategic position of India in the IOR and growing strategic importance of Indian Ocean along with increasing extra regional presence (specially China) in the region, it is time that India should term the Indian Ocean as its FRONTYARD instead of the cliché 'backyard' and start practising a stronger defence posture in the IOR with maximum coordination of national assets as well as international cooperation. Secured IOR is one of India's national objectives and the IAF along with other services is capable of achieving this objective.

To have a secured Indian Ocean with India in the captain's seat, it is necessary to predict, assess and act (not react) in terms of policies, international cooperation and better management of defence capabilities and assets. For the latter, optimum procurement and utilisation of all air power assets by the IAF should be encouraged in order to complement the growing naval capabilities and to achieve required capacity to have desired degree of control of air over the most important ocean responsible for India's growth.

NOTES

1. N.A.K. Browne, Foreword, Basic Doctrine of the IAF, 2012.
2. "The Indian Air Force in Congo", http://www.bharat-rakshak.com/IAF/History.html. (Accessed 12 February 2019).
3. "Overseas Operations of the Indian Air Force", http://www.bharatdefence kavach.com/news/indian-air-force/overseas-operations-of-the-indian-air-force/37711.html (Accessed 12 February 2019).
4. "IAF in UN Missions", https://indianairforce.nic.in/content/indian-aviation-contingent-i-iac-i (Accessed 15 February 2019).
5. N.M. Dhavle, "Chetak Detachment to Batticoloa", http://www.bharat-rakshak.com/IAF/History/1970s/1113-Dhavle.html (Accessed 13 February 2019).
6. "The Sri Lankan Interlude 1987-90", http://www.bharat-rakshak.com/IAF/History/1987IPKF.html (Accessed 13 February 2019).
7. Ibid.
8. "Operation Cactus", https://indianairforce.nic.in/op-cactus (Accessed 15 February 2019).
9. "IAF Contingent in Somalia", https://indianairforce.nic.in/content/somalia, (Accessed 15 February 2019).
10. Ibid.
11. "Indian Air Force Contingent Leaves for Sudan", 18 January 2010, https://www.airforce-technology.com/news/news74514-html/ (Accessed 13 February 2019).
12. "PM Modi hails South Sudan evacuation operation", *The Economic Times*, 17 July 2016, https://economictimes.indiatimes.com/news/politics-and-nation/pm-modi-hails-south-sudan-evacuation-operation/articleshow/53225885.cms?utm_source=contentofinterest&utm_medium=text&utm_campaign=cppst, (Accessed 14 February 2019).
13. "Aid to Civil Power", https://indianairforce.nic.in/content/aid-civil-power (Accessed 13 February 2019).
14. Ibid.
15. Ibid.
16. Ibid.
17. Ibid.
18. "Operation Insaniyat", Press Release on 14 September 2017, https://mea.gov.in/pressreleases.htm?dtl/28944/ Operation_Insaniyat__Humanitarian_ assistance_to_Bangladesh_on_account_of_influx_of_refugees, (Accessed 13 February 2019).
19. Paul J. Springer, editor, The History of Military Aviation, Olsen John A (Ed.), *Air Power Reborn-the Strategic Concepts of John Warden and John Boyd* (Annapolis: Maryland, Naval Institute Press, 2015).

20. Air Power in Indian Context, Basic Doctrine of the IAF, p. 4.
21. Characteristics of Air Power, Basic Doctrine of the IAF, pp. 18-20.
22. Gopal Suri, "Diplomatic Role for the Indian Navy and the Coast Guard", Yogendra Kumar (ed.), *Whither Indian Ocean Maritime Order* (Delhi: K W Publisher, 2017), pp. 162-164.
23. H.S. Ahluwalia, "IAF in Maritime Operations", http://bharat-rakshak.com/cms/srr/64-17.htm (Accessed 05 September 2019).
24. The Basic Doctrine of the IAF, op. cit., p. 55.
25. Ibid., p. 66.
26. Ibid. pp. 40-41, The Basic Doctrine of the IAF defines three distinct degrees of control of the air: **Air Supremacy** exists, if the enemy air power has been incapacitated to the extent that it is incapable of any air interference. This is characterised as: firstly, not being limited by time and space and secondly, being representative of the highest degree of 'control of air.' **Air Superiority** can be defined as a high degree of dominance in air, which permits the conduct of land, sea and air operations at a given time and place without prohibitive interference from the enemy air force. This condition is said to exist when, firstly, aircraft of all types can operate in all types of roles at a given time and place without serious interference from the enemy and secondly, it can limit the enemy's air operations. Further, this provides the additional advantage of information superiority due to own ability to execute desired reconnaissance missions, while preventing the enemy from doing so. **Favourable Air Situation** is limited by time and space to a much greater extent and it assures a lower degree of control of the air. Therefore, a higher degree of enemy air interference can be expected. The level of control of air is such that it enables the exercise of specific capabilities or conduct specific operations.
27. Jasjit Singh, "National Security and Aircraft Carriers", in Bhaskar C Uday, Upadhyaya Shishir (Eds.), *The Aircraft Carrier in the 21st Century* (New Delhi: KW Publishers, 2011), p. 136.
28. Anil Chopra, "Aircraft Carrier Vs Shore Based Aircraft", *South Asia Defence Strategic Review*, Vol. 13, No. 4 (September-October 2019). It is debated that today aircraft carriers may not be the most effective way to exert control over the oceans. The Department of Defence (DoD) of the US acknowledges that a cruise missile hit on a carrier, even if did not sink, will be politically and psychologically catastrophic. Same will be true for India. A powerful submarine force could be much more effective in naval warfare.
29. Ibid. The approximate cost of a medium size aircraft carrier is 6 billion USD. The aircraft and weapons onboard would be extra. For eastern IOR if this much money can be used in making ANC as an unsinkable fortress we might need only one carrier for western IOR. Another calculation says that in that much budget India can procure nearly 120 fighter aircraft like Su-30 MKI (52 million USD).
30. World Air Forces 2020, Flight Global in association with Ruag, https://www.flightglobal.com/reports/world-air-forces-2020/135665.article (Accessed 22 September 2019), p. 18.
31. Colin S. Gray, "Air Power for Strategic Effect" (Alabama: USA, Air University Press, 2014), p. 295.

32. Puneet Bhalla, "C-17: Enhancing India's Strategic Airlift capability", *Scholar Warrior*, CLAWS (Autumn 2013), pp. 100-104.
33. "Modernisation Projects of the Indian Air Force", Article in Indian Defence Yearbook (2011), pp. 290-291.
34. "C-130 J-Specification and Performance", http://www.globalsecurity.org/military/systems/aircraft/c-130j-specs.htm (Accessed 16 December 2019).
35. Basic Doctrine of the IAF, pp. 91-92.
36. Ibid., pp. 85-87.
37. Air Chief Marshal RKS Bhadauria, Chief of Air Staff, IAF, Raisina Dialogue 2020, 15 January 2020.
38. Ibid.
39. Sivapriyan, "Sukhoi-30 MKI jets inducted in Thanjavur air base", https://www.deccanherald.com/national/south/sukhoi-30-mki-jets-inducted-in-thanjavur-air-base-796480.html (Accessed 21 January 2020).
40. Press Information Bureau, Government of India, "INS 'Baaz' – Commissioned by Admiral Nirmal Verma as First Naval Air Station in Nicobar Group of Islands", 31 July 2012.
41. P.K. Roy and Cawasji Aspi, *Strategic Vision 2030: Security and Development of ANC* (New Delhi: Vij Books, 2017), p. 86.
42. Ibid., p. 155.
43. Panda Ankit, "South China Sea: What China's First Strategic Bomber Landing on Woody Island Means", *The Diplomat*, 22 May 2018.
44. Roy & Cawasji, op. cit., p. 87.
45. "How CDS can integrate the services", *The Sunday Tribune*, 29 December 2019.
46. Basic Doctrine of the IAF, op. cit. p. 67.
47. Population of Overseas Indians, https://mea.gov.in/images/attach/NRIs-and-PIOs_1.pdf (Accessed 20 June 2020).
48. B.S. Dhanoa, "Air Ambulance landmark achievement in trauma surgery", 25 June 2017, https://www.financialexpress.com/india-news/air-ambulance-landmark-achievement-in-trauma-surgery-indian-air-force-chief/735351/, (Accessed 28 November 2019).
49. "Updates: How Navy Commander Abhilash Tomy was rescued", https://www.indiatoday.in/india/story/live-updates-abhilash-tomy-stranded-indian-navy-officer-rescued-safely-1347408-2018-09-24 (Accessed 28 November 2019).
50. Basic Doctrine of the IAF, op. cit., p. 116.
51. LIMA 2019, https://indianairforce.nic.in/content/lima-2019 (Accessed 15 July 2019).
52. ANI, "Defence Space agency to come up at Bengaluru", 15 May 2019, https://economictimes.indiatimes.com/news/defence/defence-space-agency-to-come-up-at-bengaluru/articleshow/69345697.cms?from=mdr (Accessed 20 July 2019).
53. Indian Regional Navigation Satellite System (IRNSS) : NavIC, https://www.isro.gov.in/irnss-programme (Accessed 02 December 2019).
54. Rajeswari Pillai and Nagendra Narayan Prasad, "Creation of a defence space agency: A new chapter in exploring India's space security", https://

www.orfonline.org/expert-speak/creation-of-a-defence-space-agency-a-new-chapter-in-exploring-indias-space-security/ (Accessed 02 December 2019).

55. "Towards Self Reliance in Navigation-IRNSS", https://www.isro.gov.in/towards-self-reliance-navigation-irnss (Accessed 02 December 2019). According to Indian Space Research Organisation (ISRO), IRNSS system is expected to provide position accuracy of better than 20 m over Indian region and an area extending about 1500 sq km around India. IRNSS constellation consists of three Geostationary Earth Orbit (GEO) spacecraft and four spacecraft in geosynchronous orbit inclined at 29 deg. It would provide two types of services, namely, Standard Positioning Services (SPS) – provided to all users – and Restricted Services (RS), provided to authorize users.
56. Brilliantwiki2, https://commons.wikimedia.org/wiki/File:IRNSS_ Coverage_Polygon.png#file (Accessed 02 December 2019).
57. Cristina Goulter and Pant, Harsh V, "Realignment and Indian air power doctrine", *Journal of Indo-Pacific Affairs* (Fall 2018), p. 31.
58. Ibid.
59. https://mea.gov.in/circulars-notifications.htm (Accessed 30 December 2019).
60. https://mea.gov.in/divisions.htm (webpage of MEA divisions was browsed to make the list and placed as appendix 2) (Accessed 30 December 2019).
61. Personal interview with Mr Amit Narang, JS IOR, MEA on 16 December 2019.
62. Ahluwalia, op. cit.

7

Recommendations

The Government of India is cognisant of the importance of a secured IOR. Diplomatic engagements have increased and there is positive response from many nations. China's overtures in the IOR have caught everyone's attention which has led to one common opinion that China's strategic ambitions should not be underestimated. But one should also not overestimate China and lose focus on the principal of mutual growth. India has always respected the sovereignty of all its neighbours while assisting them in growth so as to have a strong and reliable neighbourhood. This approach of India is articulated well in India's 'Neighbourhood first', 'Act East' policies and now through Prime Minister's SAGAR vision, it is clear that India believes in making the IOR a secured place for mutual growth of all IOR nations while safeguarding own national interests.

Few recommendations where the IAF can contribute towards this vision and add to the enhanced security of the nation as well as the IOR by spearheading the holistic use of air power are as follows:

1. HADR Missions

- A well planned procedure for HADR oriented exercises should be put in place which includes reaching far flung islands of IOR countries by air and landing at different runways there with simulated relief material or/and rescue teams. A joint survey also should take place of the topography of vulnerable areas and then joint drills with the domestic organisations involved in disaster relief will only increase the ease in communication and execution while reducing wastage of the most critical factor – time. Such survey and analysis oriented visits will build trust factor and establish positive outlook towards India.

- Coordination between medical agencies is also important to provide immediate rescue and relief. Few IOR countries like Maldives have chain of many islands which are distant from each other. Lack of connectivity as well as medical services at all islands may cause avoidable fatalities. Medical teams of the IAF can be deputed after mutual agreement and the IAF aircraft in CCATT configuration can be given the task of air ambulances for defined periods. If required, aircraft in air ambulance configuration can be given on lease as well, to be used by Maldivian National Defence Forces (MNDF) or respective agency of any country which needs such assistance.

- The capability of search and rescue through joint operations using the IAF transport aircraft and other elements of the Navy and the Army should be tried in practice drills. It will be useful in contingencies like medical emergencies in the high seas (case of Abhilash Tomy[1]) as well as rescuing hijacked ship or to assist a capsized ship's crew.

2. **Non-Combatant Evacuation (NCE) Missions**

- Given various situations like recent Corona virus outbreak, civil unrest in Yemen etc., frequent exercises and drills simulating contingencies which can cause large evacuation missions, should be conducted including operations by the IAF aircraft with heavy lift capabilities. The success of future operations will also rely on New Delhi's willingness to work together with friendly governments. India will have to invest in cooperative frameworks that facilitate coordination among countries that have large expatriate populations in West Asia, in particular Nepal, Bangladesh and Sri Lanka and among leading powers with evacuation capacity in the Indian Ocean region.[2]
- The government will have to assign a greater role to its defence forces, in particular by strengthening the Navy and the Air Force's capacity to operate in tandem with civilian authorities. It should direct the military to develop a NCE doctrine, designate the Integrated Defence Staff as the nodal organisation to improve existing inter-services and civil-military coordination, direct the services to conduct more multilateral NCE exercises, and adapt military modernisation plans to increase capacity for out-of-area deployment and evacuation.[3]
- It is also recommended that the OOAC policy/doctrine should be issued with a follow-up plan indicating responsibilities and a time-line for activation. An inter-ministerial committee should prepare a manual with guidelines that establish a clear chain of command and division of competencies; identify regional support bases, assembly points and routes for evacuation; develop country-specific warden systems to communicate with expatriates;

and establish evacuation priority and embarkation criteria.[4] The MEA may look at setting up an Emergency Division with provision of having a team consisting of participants from all agencies like Defence, Finance, Health, Home, and Civil Aviation Ministries to cater for such situations.

3. **Tapping Strategic Potential of Islands**: The ever increasing strategic importance of Indian Ocean, especially in terms of power demonstration by extra regional countries demands overall development including basing of a strong force level at India's island territories.[5] Associated infrastructure development should focus on:

    - Upgrading of Car Nicobar runway with an option of developing it as international military-cum-civil airfield because it can be a good air link between Southern and Northern islands. This will also help in decongesting the Port Blair airport. If international flights are allowed to land here directly it will be a very good island destination for many South East Asian countries. It will also serve as nearby diversion for civil flights which presently take Chennai or Kolkata (which are miles away) as diversion to Port Blair airport.

    - Extension of runways at Campbell Bay and Shibpur for heavier aircraft. INS Baaz at Campbell Bay has one runway of inadequate length for all types of aircraft and extension seems difficult. The option of acquiring land for the IAF and developing a longer runway can be exploited. Reclamation of land from sea is another technology which India is capable of and this option should be looked into after thorough consultation with authorities responsible for ecological conservation.

- Enhancement of existing infrastructure set up at the defence land for air operations at Great Nicobar.
- Infrastructure enhancement to facilitate fighter operations from Port Blair in case of contingencies. Frequency of fighter aircraft operations to Car Nicobar should be increased to give exposure of extended and joint maritime operations to pilots of the IAF. Once the squadron strength of the IAF increases, option of having a semi-permanent detachment of fighters can be looked into.
- To have a good AD network, a fair number of High and Medium Power Radars and SAMs should be placed at carefully surveyed location along the length of the archipelago to cater for possible air threat emerging from any direction. Development of a SAM test firing range shall be looked into so that necessary support infrastructure comes up for missile units of the IAF.
- Development of existing air link and infrastructure to support air power operations from Lakshadweep islands is very important. Extension of Agatti runway and infrastructure setup will enable heavier aircraft with more passenger/cargo load to operate. An airfield at Minicoy with necessary infrastructure must be given priority.
- To exploit the full potential of existing military set up under ANC, force structure needs to be reviewed while enhancing jointness between service headquarters and developing present infrastructure. Lack of jointness in operations, logistics, training, support services, communications, repairs and maintenance needs attention. Option of replacing ANC with IOR Command with the IAF having complete responsibility of air cover and air assets may be looked into.

4. The Navy has berthing rights at many international ports. It includes logistic support like refuelling, repair etc. for the Indian Navy ships at harbours of other countries. On similar lines, landing rights for the IAF aircraft also should be obtained as part of defence agreements or MoU. The option of using airfields on island territories of friendly countries in the IOR for refuelling halts during ferries, HADR exercises/ contingencies or conducting joint exercises with navy or air force of host countries must be looked into. It will give exposure to the IAF pilots for prolonged operations over sea and landing at foreign bases, which will help in future OOAC and HADR operations. There are dependencies and islands of other countries in the IOR with suitable infrastructure for air operations. Cocos Keeling and Christian Island of Australia are such example. India may reciprocate by conducting joint air and naval exercises at Andaman Islands and giving logistic support to selected countries for peacetime operations.

5. **Multilateral Platform with IOR Countries for Air Policing Regional Initiative:** Though it is good to have bilateral exercises and defence cooperation but to have a more stable safeguarding measure, the option to initiate brainstorming for a regional platform focusing only on interaction of air forces or air wings of the IOR countries can be looked into.[6] This platform can look into using air and space assets to do tasks of air patrolling, surveillance of the EEZs, HADR operations, protection against piracy, hijacking, appropriate actions against sea borne identified threats, providing immediate medical assistance etc. The joint exercises would be conducted on similar themes like sharing of radar pictures which in future can solve the problems like disappearance of flight MH-370 or tackling a 9/11 type of aerial threat. Extended ADIZs with mutual consent should be created with purpose of having

enhanced AD cover over critical parts of the IOR with the IAF being the lead agency. Flight safety oriented discussions, joint accident investigations and aerospace management interactions will optimise the purpose of such a platform.[7] If such an air symposium is formed, all the countries will be on a mutual platform which will assure assistance in any eventuality which can ultimately be a threat to peace and growth of the IOR.

6. **Integrated Air Power for Maritime Security:** The air assets of the IAF and the Navy can jointly be used towards the larger goal of providing overall security to the IOR. This concept will require 'joint planning of operations and not the planning of joint operations'. It will reduce the redundancy of efforts for a specific mission or task and also will cut down the cost of operations carried out by individual services.[8]

   Idea of doing joint exercises with the Navy on regular basis shall be exploited (not only in capacity of observers on board). Exercises including AWACS, FRA, P-8I, Su 30s, Mirage 2000s, MiG 29Ks, Jaguars, CBG, UAVs should be conducted in in the AOR and with further tie ups, refuelling halts also should be planned at friendly IOR countries. This will bring in direct exposure and more streamlining to the concept of applying integrated air power in the IOR.

7. **Joint Procurement of Force Multipliers:** The assets which will help enhance air power in maritime domain shall be procured through joint planning. Strategic assets like AWACS, FRAs, LRMP aircraft, UCAVs etc. are such platforms which should be procured after detailed study of area of operations, future requirements and present shortage. The procurement process should also focus on common platforms like fighter aircraft or helicopters, so as to obtain economically sound deals from common manufacturer or supplier. Strategic bombers and

nuclear submarines are also force multipliers which can be looked at for procurement as joint strategic assets. Investment on technology that can enhance the capabilities of existing platforms is also equally important. For example, while IACCS is successful, networking and communication shortcomings are still hindering the IAF from becoming a complete networked force with the Army and the Navy.[9]

8. Global and regional affairs specialist from defence forces to be sent to countries of interest in the IOR. Other than update through defence attaches, there should be short term deputations/visits of knowledgeable scholars from three services to interact with the respective branches in those countries. The purpose is to have a deeper knowledge exchange bonding and to know more about the specific needs of the IOR nations which can be met through active interaction by Indian defence forces. Generating 'scholars in uniform' with good regional knowledge through a systematic programme should take priority in the larger interest of the country.[10]

9. There is a need to increase opportunities for the IAF to interact with other air forces/air wings of the IOR for capacity building in terms of training, equipment as well as maintenance. India can help friendly IOR countries develop their air wings, training facilities and maintenance practices after thorough diplomatic discussions on such topics. Dedicated attention on conducting the IAF training detachments on semi-permanent basis with the IAF personnel, aircraft and equipment shall enhance India's soft power presence. Many countries operate same aircraft as the IAF e.g. Mirage by UAE, Rafale by Qatar, MiG 29 by Myanmar and Su-30 by Indonesia, Malaysia and Vietnam. This itself creates an opportunity for training exchanges.

10. Selected officers of the IAF must be sent to various divisions of MEA on deputation. They should be selected on basis of their

knowledge in international affairs and their performance in staff courses (subject of area studies). Officers who are doing or have done some research specific to a zone/country should be encouraged. It will ensure smooth amalgamation of defence and foreign policy matters.

11. UN peace keeping mission cell should be shifted from Army HQ to HQ, IDS which shall do the centralised detailing of personnel and assets. It will ensure active involvement of all services as per their resource availability and will represent jointness of Indian Armed Forces in terms of delegation to such missions.

12. There should be a dedicated directorate for international cooperation in the Air HQ with an air officer as in-charge. The existing sub directorate can be changed into a specific directorate of foreign cooperation with dedicated region wise officers to handle interactions, engagements and deputations. Additional staff can be taken from existing directorates like from project directorate for export promotion, from training directorate for training exchanges, from operations directorate for international exercises etc.

13. The IAF should enhance its OOAC capability in the IOR. Since directive of the CDS includes emphasis on HADR and OOAC capability of defence forces, missions and exercises on these themes with other countries should be conducted on frequent basis. The IAF should not wait for HADR contingency to occur and then scramble. Indian flag on tails of C-17s and C-130s of the IAF should become a normal sight in the IOR countries, so that they are used to India's friendly presence and mutual respect is generated. Sudden appearance will create doubts. It will make pilots of the IAF also more aware of the terrain, defence culture, requirements, mutual strengths and weaknesses.

14. The IAF has been successfully upgrading many airfields under MAFI project. Under this project the IAF has upgraded the navigational aids and associated airfield infrastructure to enhance the operational capability by facilitating air operations of military and civil aircraft even in poor visibility conditions/adverse weather[11] while enhancing aerospace safety. India can promote this project in IOR countries which are willing to upgrade their airfields. Technical help during set up and training can be given by experienced personnel from the IAF.

15. Setting up AD radars on strategically important islands in the IOR is necessary so that first principle of air defence – detection is achieved. The radar pictures can be shared with operational centre of Southern Air Command and HQ MAO to have a clear situational awareness and larger view of the region. India is helping few countries in setting up Coastal Surveillance Radar Stations (CSRS) for better awareness about maritime traffic. Sharing of MDA about vessels in the Indian Ocean through IFC-IOR at IMAC in India has been welcomed by many countries. The air awareness is equally important and can be achieved by integrating the air picture picked by sensors located at various islands.

16. Satellite based surveillance needs to increase while enhancing Indian space capability. To have a robust C^4ISR system suitable for network-centric operations, more ISR dedicated satellites and their integration in an overall improved intelligence apparatus must be given priority.

17. **Regional Navigation Satellite System:** There is a need to expedite setting up the ground support system for IRNSS. Once the complete system of indigenous navigational services is tested for military and civil purposes, other IOR countries shall be assured of secured navigation throughout the region.

18. **Akashmala Project:** Development of ports in other IOR countries with Indian help is being considered already. Indian government's Sagarmala project to develop and connect inland waterways within the country and neighbouring East Asian countries is picking up good response. A project with name of 'Akashmala' for airport infrastructure development in friendly IOR countries should be looked into as per their requirement. India has good private companies to undertake such work. It will boost economic ties as well as create employment opportunities. If Indian government offers to undertake the maintenance of such airfields, then it eases out the overall cost burden on the country. India has one such base in Tajikistan which is maintained by both the IAF and the Air Force of Tajikistan. Mauritius, Mozambique, Socotra (Yemen), Madagascar, Maldives are potential nations which can be actively approached for improvement of air operations with help of the IAF.

19. **Defence Cooperation Policy:** Currently, the defence cooperation engagement plan, is drawn up by the Services and executed after clearance from the Cabinet Secretary (where applicable), the MEA and the MoD, thus making it rather a 'bottom up' than a preferred 'top down' approach. In order to exploit the full potential of the tool of defence cooperation, it is vital that various agencies involved with the conceptualisation of the defence cooperation events like the MEA, the MoD and the Department of Military Affairs (DMA), evolve a mutually consulted both long term (two to five years roll on plan) and short term (annual) defence cooperation plan so that these are implemented smoothly and with minimal slippages.[12]

20. **Active Interaction through Think Tanks:** Active dialogues between the likeminded nations on various issues on regular basis are very important. It is not always possible to have

meetings between delegates and ministers or exchange between services and bureaucrats. That's where think tank community can bridge the gap by interacting with other countries' think tanks. Maritime security involving all elements of defence forces should be the theme of conferences conducted by think tanks across IOR countries. For example, India Foundation, a prominent think tank in New Delhi, conducts Indian Ocean Conference (IOC) every year. The IOC is an alternative platform where the participating nations come together to exchange views on various issues pertaining to the region's maritime security at large. Even though there are other institutional mechanisms available for the parties concerned to address the same issues, it is a good sign that this annual conference is taking its own institutional shape with each passing year, and is attracting participation from more and more countries around the world.[13] Similarly defence think tanks of India shall engage with counterparts in the countries of the IOR on a regular basis with active participation from all three services.

21. The IAF sports team is winning many laurels across the world in many streams. The sports interactions on side-lines of diplomatic engagements with IOR countries can build trust factor and improve mutual understanding in defence forces.

22. **Air Shows:** The IAF can make the air shows a good platform of advancing Indian diplomatic and commercial interests. While conducting and participating in important air shows of the region, the IAF under its senior leadership should promote Indian defence industry and with its professional aerial displays gather more appreciations for the country.

NOTES

1. Cf ante, Chapter 6, p. 100.
2. Constantino Xavier, "Getting Back Home, Safely", https://carnegieindia.org/

2017/02/08/getting-back-home-safely-pub-67974 (Accessed 12 November 2019). His article about bringing Indians back safely, mentions that India has extensive experience in conducting evacuation operations, but to secure the lives and assets of Indians abroad, the government must avoid an ad hoc approach and seek to institutionalize best practices, bolster diplomatic and military capabilities, and improve coordination. He also suggests that Studying India's history, best practices and lessons learned will help institutionalise them and avoid the need to reinvent the wheel every time a crisis erupts. By supporting policy-oriented research at universities and think tanks to document the memory of senior officials, the government would also facilitate the transmission of their expertise to younger officials. Other recommendations mentioned are: (a) to minimise redundancies, the government must institutionalise a permanent inter-ministerial coordinating mechanism for emergency evacuations, incentivise inter-agency cross-posting of officials dealing with diaspora affairs, and encourage State governments to create regional contingency plans, (b) to avoid cost inflation and delays, the government must establish a permanent civil reserve air fleet that pools aircraft from all Indian airlines based on pre-established requisition and reimbursement procedures, (c) India's diplomatic cadre must be given specific training to operate in hostile environments. To achieve this, the government could instruct the police or army to train Indian Foreign Service probationers to operate in war zones; conduct frequent evacuation simulations and emergency drills; and create rapid reaction teams of Indian security personnel to be deployed to protect diplomatic staff and installations abroad.

3. Ibid.
4. Ibid.
5. P.K. Roy and Cawasji Aspi, *Strategic Vision 2030: Security and Development of ANC* (New Delhi: Vij Books, 2017), p. 157.
6. To have a secured environment in the IOR, a consensus on security architecture is a must. This consensus cannot be imposed or generated. It has to come through discussion, analysis of present situations and mutual understanding of all members; and also evaluation of present security architectures. IONS, IORA, BIMSTEC are few systems in place but their purpose is based on either economic cooperation or at best to conduct few defence exercises along with joint/coordinated patrolling of the EEZs which is not based on a multilateral understanding.
7. The participation by the countries would depend on their asset availability and willingness. Few counties are not having dedicated air wings or air forces, however they can induct their special forces, protection forces etc for joint training as it will be helpful in case of actual requirement. All the countries willing to participate in such symposium should be sharing the scale of effort, resources and time line to such joint exercises so that it can be planned on regular basis.
8. With the creation of the post of CDS formally on 31 December 2019, the concept of integrating national defence assets is likely to shift place from books to reality. For maritime security of the country and the IOR, it is imperative that all the national defence assets have to be used in an integrated fashion. The air power

assets are best utilised under command of air minded experts and maritime assets by experienced naval experts.

9. Angad Singh, "Focus Beyond Fighters", https://www.orfonline.org/expert-speak/focus-beyond-fighters-56374/ (Accessed 26 December 2019).
10. Vikas Kalyani, "Global Affairs Expert Programme", *Salute*, Vol. 11, No. 2 (October-November 2018), p. 26.
11. "Vice Chief of the Air Staff Commissions Pilot Project on Modernisation of Airfield Infrastructure (MAFI)", 26 March 2014, https://pib.gov.in/newsite/PrintRelease.aspx?relid=104691.
12. Arvind Dutta, "Role of India's Defence Cooperation Initiatives in Meeting the Foreign Policy Goals", *Journal of Defence Studies*, Vol. 3, No. 3 (July 2009), p. 33.
13. Martand Jha, "Why the Indian Ocean Conference is Becoming Important", https://www.telegraphindia.com/opinion/why-the-indian-ocean-conference-is-becoming-important/cid/ 1670382.

8

Conclusion

Indian Ocean has a unique history. It has been a major channel of cultures, trade, commerce and religions much before other oceans. Concept of globalisation began in the Indian Ocean littoral about 4,000 years ago and it whittled the shape of the region what is present day IOR. It connected India to many countries that were dependent on trade winds and established India as the strongest economy of the world till few centuries back. Undoubtedly IOR and India are in a way complementary. IOR was in fact a self-sustained, free and interconnected world with India as the pivot because of location and cultural and economic weight. Much before Vasco da Gama reached western coast of India, Arab sailors were taking halts there in their journey to South East Asian countries. After many historical events, today the IOR has regained its importance and is the most strategic economic corridor supporting major economies of the world.

The complexities of current geo-political affairs in the present world are making it difficult to gauge the tumults in the IOR of future. The fragile sensitivities of Gulf along with disturbed AF-Pak region can become the reason of a disturbed Western IOR. On the other hand, the spill over of turbulences from the SCS in the Eastern IOR can make waters of Bay of Bengal choppy. In this wide region ranging from Egypt to Australia, a universally established coherent vision about the security challenges has been lacking since the British Navy's withdrawal. Presently the military balance in the IOR is largely underpinned by the US Navy but Chinese and Russian presence is also increasing. The US, British and French governments are speculating the worsening future competition of major powers in the IOR and showing their faith in Indian government's capability to play a leading role for security balance in the region. Japan and Australia are also of the same opinion as is evident from increasing defence dialogues between India and these countries. At this juncture, the IOR needs a balanced coalition of littoral nations, a platform for understanding, an arrangement for secured and assured growth in an environment of peaceful coexistence while catering for future digressions and applicable solutions. Because the primary responsibility for peace, stability and prosperity lies on the countries of the Indian Ocean littoral and not on extra regional powers. India is conscious of its significantly important geostrategic location in the IOR and not only for her own secured present but also for a peaceful and progressive future of the region, India is developing strategic ties with littoral counties.

SAGAR is a comprehensive vision for inclusive development for all Indian Ocean nations, if adopted in true sense. It is a guiding principle that can lead to India's growth as well as help in generating respectful relations with IOR nations for mutual development in a safe and secure environment. The basic tenet of this vision is to deepen economic and security cooperation with maritime neighbours and enhance overall maritime security environment and economic strength. It envisages collective action and cooperation to advance peace and security and response to emergencies through multilateral diplomacy and bilateral cooperation. It also seeks a more integrated and cooperative understanding for the region that focuses on sustainable development for all. This vision needs well-coordinated implementation within the country as well as with other countries in the IOR. Besides a potential foreign policy promoter, SAGAR should be treated as extension of India's defence policy as well because for any blue water economy to survive it is imperative to have a strong formidable maritime defence policy and good international defence cooperation podiums and this vision has the potential to provide the IOR such platform.

For ensuring the security of the country and the IOR under this vision, the defence forces of India will play an instrumental role. The three services are experienced and seasoned enough to promote the task of nation building across the oceans. Peace keeping operations, anti-piracy operations and drills, various HADR missions and NCE missions are examples of contribution by the defence forces towards a stable and secure IOR. Over 88 years of illustrious existence, the IAF has attained the confidence for carrying out all kinds of operations including operations out of the area. The directive of recently appointed CDS also includes enhancing India's capabilities to conduct these out of area operations as well as HADR missions. Other than countless rescue missions within the country, many nations have witnessed the HADR and NCE capabilities of the IAF. There are other examples also in history when IAF undertook combat missions to support another country and successfully played the role of a situation stabiliser. IAF is currently one of the most experienced forces to act quickly and play vital role in situations like natural disasters, unconventional security threat of limited intensity or even conventional conflicts. Hence it is an accomplished force to safeguard national and regional interests under guidance of the CAS, as and when directed by the GoI. By using characteristics of air power like mobility in third dimension, speed, reach, responsiveness and flexibility in most suitable ways, the IAF possesses the ability of providing the IOR countries required assistance, support and cooperation to overcome all kinds of challenges which pose threat to growth and security of the region.

With the present capabilities, the IAF has been, through its performance in international exercises, gaining prominent respect from major powers of the world. The number of such engagements is lowest for the IAF but is increasing gradually. IOR countries are at bottom of list of countries with which the IAF has engaged in any sort of international defence cooperation and this aspect needs to be looked into. Other than conducting joint exercises there are many capacity building measures that India can take up through the IAF like, providing training to all branches of other air forces, providing exposure to modern indigenous equipment for further procurement through appropriate channels, helping in set up of defence infrastructure related to aerospace power, frequent specialist exchanges to build better understanding and conducting air shows on important events of individual country willing to host the IAF for magnificent display of courage and coordination.

It is time that India should use vision SAGAR to strengthen its maritime defence policy as a regional defence policy with active inclusion of its air power assets and with maximum IOR countries on-board. India then, shall be able to assume the role of regional security provider. To have a secured Indian Ocean with India in the lead role, it is necessary to do three Es – Envisage, Evaluate and Enact (not react) in terms of policies, international cooperation and better management of defence capabilities and well planned procurement of assets. For this, careful consideration should be given to procurement of critical aviation assets such as FRA, AWACS etc. that can enhance strategic reach and efficiency of air power operations along with investing in building net centricity for such operations across the IOR. Strengthening and augmenting infrastructure at its natural strategic outposts of island chains of

Andaman & Nicobar Islands and Lakshadweep is also an important step in India's Indian Ocean policy. By developing more dual use airfields, fit for operation of heavier transport aircraft and associated infrastructure, not only the air-link for international tourists can get better but also the upkeep of various neglected islands will improve. Enhancing structure of present joint command at ANI and giving it more teeth in terms of assets, infra and responsibilities will only enhance the core theme of SAGAR i.e. security and growth of IOR.

With the aim of creating a secured environment around Indian maritime domain as well as around other friendly IOR nations, the defence posture in the IOR needs to be improved by enhancing air power practices amongst the nations of the region. While India respects and follows the principal of freedom of navigation and abides by UNCLOS, the responsible posture by India will ensure that there is no violation of rules and no challenges to sovereignty, assets, SLOCs and peace of any of the IOR countries. India needs to think rationally and act responsibly to secure the position of leading nation for overall growth of the IOR, the one it had for thousands of years in past.

India has to focus on its capability to make the IOR a self-sufficient, sustainable, peaceful and secured region. It should use its assets, policies and relations in such a way that the world can see the IOR as a homogenous entity with diversity and yet united to grow together under leadership of India. The SAGAR vision for Hind Mahasagar (Indian Ocean in Hindi) can in fact be expanded as MAHASAGAR vision – Mutual, Aggregate, Homogenous and Assured Safety And Growth for All in the Region.

9

Addendum: COVID-19 and Infected Geo-politics

The year 2020 started on a normal note like every other year, but the coming months were to introduce the world with a pandemic called COVID-19, origins of which were found in December 2019, hence the name. Initially, there were some reports of a pneumonia-like disease in Wuhan, China. A few days later it was revealed that a new type of Corona virus was causing the disease and eventual deaths, which quickly rose to concerning numbers.[1] With no existing therapy working on the virus, rapid spread in many countries made it soon an international crisis. The place of origin, as per many reports, has been established as Wuhan Institute of Virology, a laboratory in Wuhan, the capital city of Hubei province in central China. A lockdown was imposed in Wuhan on 23 January 2020. About one and a half month later, the WHO declared spread of Corona virus as a global pandemic on 11 March 2020.[2]

COVID-19 has emerged as a very dangerous pandemic which has affected all countries especially western and European economies.[3] As on 30 October 2020, the number of cases worldwide were 46,384,390 and 1,199,896 deaths due to this virus had taken place.[4] Fortunately, the overall situation in India is far better as compared to most of the other major economies. By the same date, India had faced approximately 8,136,166 cases and 121,681 deaths[5] with an appreciable recovery rate nearing 92 per cent.[6] India's largesse and preventive measures against this global menace have been being praised by the global community, including the WHO.[7]

Impact on World Economy

The pandemic, which was first detected in China in December 2019, had infected people in 188 countries by June 2020 in a short span of six months. Its spread left businesses around the world counting costs and vulnerable to huge losses. The IMF predicted the global economy to shrink by 3 per cent in 2020. It described the decline as the worst since the Great Depression of the 1930s.

Fig. 9.1: Majority of Countries on the Brink of Recession
Real GDP growth, Q1 2020

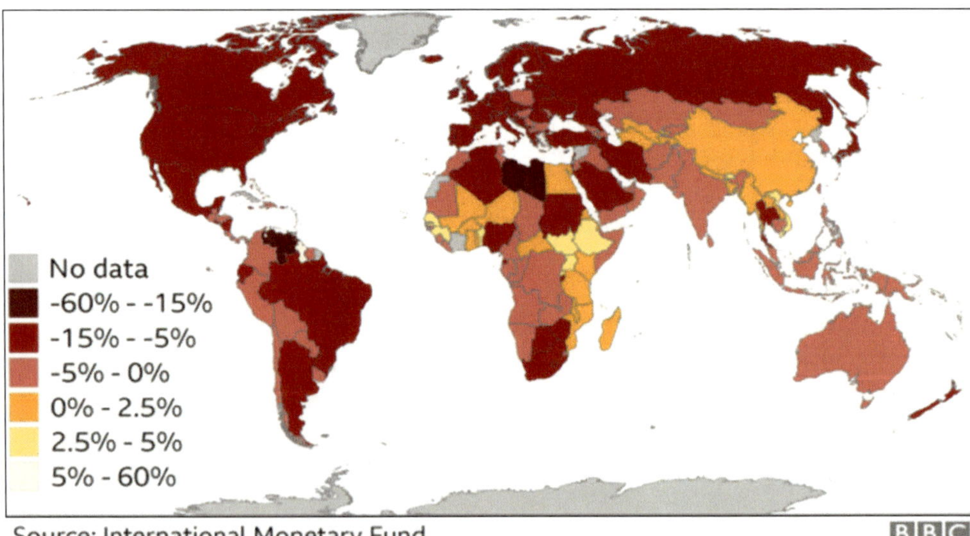

Although the IMF said that the coronavirus has plunged the world into a "crisis like no other", it had expected global growth to rise to 5.8 per cent in 2021 if the pandemic had faded in the second half of 2020. That's driven primarily by growth in countries such as India and China. Recovery in big, services-reliant economies that have been hit hard by the outbreak, such as the UK or Italy, is expected to be a slow process.[8] Due to the fall in travel, global industrial activity has been affected. Oil prices fell further in March 2020 as the transportation sector, which accounts for 60 per cent of the oil demand, was hit due to several countries imposing lockdowns.[9] The price of crude oil dropped below USD 20, which was lowest level seen in 18 years. However, crude oil prices doubled by November 2020 since their April low, supported by sharp oil supply cuts, but prices remain one-third lower than pre-pandemic. Metal prices recovered rapidly in response to a faster-than-expected pick up of China's industrial activity.[10] When most economies were shrinking, the Chinese economy claimed to recover in the second quarter (3.2 per cent growth) and shown growth of 4.9 percent in the third quarter[11] whereas, the GDP in India was reported by some statistics to have contracted 23.9 per cent in Jun 2020, following a growth of 3.1 per cent in the next quarter.[12] That means that China only recorded one quarter of negative growth due to the pandemic.

Fig. 9.2: Coronavirus Slump Continues for GDP Growth
2020 Q3 GDP growth in selected counties (in percent)

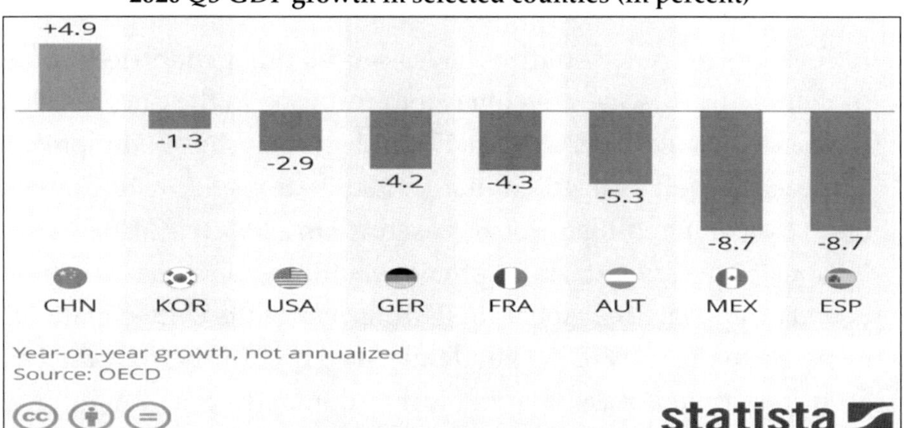

Source: https://cdn.statcdn.com/Infographic/images/normal/18095.jpeg

It may be because of a robust domestic demand and China's ability to fund infrastructure projects all over the world. However, it is too early to think that the Chinese growth trajectory is back on its continuous upward path. While the effects of COVID-19 on the global economy and geo-strategy would be more extensive than can be assessed at present, even without this global pandemic, China was actually facing some very serious challenges to its economic growth which have only grown marginally since and as a result, its erratic behaviour is quite visible.

Effect on International Relations

The COVID-19 pandemic has impacted international relations and affected the political systems of multiple countries, causing suspensions of legislative activities, isolation or deaths of multiple politicians and even rescheduling of elections due to fears of spreading the virus. The response to the pandemic has resulted in unprecedented expansion of government powers in many countries. At least 84 countries declared a state of emergency in response to the pandemic, leading to fears about misuse of power. Reportedly, 38 countries restricted freedom of the press as a result and in some parts the 'would-be autocrats' used COVID-19 as an excuse to grab more power.[13]

Leaders of some countries have accused other countries for not containing the disease effectively and resulting in the uncontrolled spread of the virus. The COVID-19 pandemic has caused diplomatic tensions, as well as resulted in a United Nations Security Council resolution demanding a global ceasefire. On 23 March 2020, United Nations Secretary-General António Manuel de Oliveira Guterres issued an appeal for a global ceasefire as part of the United Nations' response to the COVID-19 pandemic. And on 01 July 2020, the UN Security Council passed a resolution demanding a general and immediate cessation of hostilities for at least 90 days and requested the UN Secretary-General to help ensure accelerated international response to the coronavirus pandemic.[14] The call gained broad political support and encouraged hopes that the pandemic might serve as a catalyst for cessation of armed hostilities. Nevertheless, if anything, a trend of intensifying conflicts and increased insecurity has been observed, as the pandemic responses have created opportunities for armed actors.[15] Public health crisis, weak policy responses and economic fallout are three main ways in which this pandemic has affected many countries and created a gap for external players to cause unrest.

On the other hand, international and domestic travel restrictions have led to the virtual decimation of travel, tourism and related industries, apart from stranded students and tourists in foreign lands, with no means of support. In the process of repatriation of stranded citizens, host countries behaved in a particular way to countries of origin, which also caused a situation of discomfort in otherwise normal relations. Hence the international relations are going through testing waters in these tough times. The predictions regarding the 'de-globalisation' and talk of the 'Suez moment', the 'Pearl Harbour moment' and 'China's Chernobyl' have been in the air.[16] Clearly a new world order is in the making.

China in the Accused Box

According to many reports, China's giant opaque bureaucratic system did not work efficiently in containment of the virus from Wuhan. The most important issue this outbreak exposed was the local government's lackadaisical response and fear of action. As the virus spread, officials in Wuhan and around the country withheld critical information, played down the threat and rebuked doctors who tried to raise the alarm. A reconstruction of the disease's spread by The New York Times showed that by not issuing earlier warnings, the Chinese government potentially lost the window to keep the disease from becoming an epidemic[17] which later turned in pandemic.

The French Institute of International Relations (IFRI) report said that China's diplomatic program has now "gone into overdrive" as it battles accusations of allowing the novel corona virus to spread, and failing to warn the international community of the danger or be transparent about its own number of infections and deaths. Multiple lawmakers in Europe and the US have called for an international investigation into the origins of COVID-19, a proposal dismissed by Beijing.[18] The pre-COVID geopolitics driven by considerations of strategic and financial gain, was projected as "involving China as a responsible stakeholder in the global order" by many countries, has demanded a re-look. Countries such as France, Australia and Germany have joined the US in questioning the credibility of China's COVID-19 statistics and its management of the crisis.[19]

China's supposedly notorious role in manufacturing this virus in its Wuhan Lab and then possibly releasing it in the world—thereby causing unimaginable loss of precious lives and many other long lasting impacts over the emerging global scenes has put the dragon in a box. Unfortunately, China is said to be in possession of many other dangerous viruses in its aforesaid lab, and that probably puts the entire humanity at ransom to the hands of Beijing.[20]

China's Post-COVID Behaviour

If at all, the COVID crisis was to present an opportunity for hegemonic countries to assert their narratives when all other are busy recovering from the plunge or getting used to of new normal, China appears to be leading the list. Having been cornered by many economic powers and seeing the threat of economic strangulation of the great Chinese dream, China has started behaving like the other pole in its imaginary bipolar world. China's 'wolf warrior diplomacy' has given way to tense geopolitical confrontations with a growing array of its neighbours, from stand-offs with Vietnam and Taiwan in the South China Sea to threatening Australia with boycotts of wine, beef, barley, and Chinese students.[21] China's continued attempt to blackmail Canada has backfired spectacularly. When Canada arrested Meng Wanzhou, the chief financial officer of Huawei, based on a US request under the mutual extradition treaty, China retaliated by arresting two Canadians, Michael Kovrig and Michael Spavor. China was also verbally harsh with the United Kingdom when its Foreign Secretary Dominic Rabb said that, along with Canada and Australia, his country would suspend the extradition treaty with Hong Kong over concerns about the new security law. The United Kingdom also said that by 2027, all equipment from Huawei, the Chinese telecom giant, should be purged from his country's telecom sector.[22]

On the other hand, the Chinese territorial onslaught as part of its 'military coercion' strategy across the 'nine dash line' that consists strategic interests of all the nations in the South China Sea has increased in the past months. China is unremittingly claiming the maritime space and even putting on displays of air power. Taiwan's newest Coast Guard patrol vessel was damaged after being rammed by Chinese fishing boats near Kinmen County. A Vietnamese fishing boat was also attacked near the Paracel Islands, an archipelago in the South China Sea that China claims as its own.[23] China has been putting pressure on Malaysia and Indonesia as well by encroachment of fisheries and trailing their ships in the SCS. Japanese Navy also has seen few incursions and one of the US vessels was also expelled by PLAN. PLAN further flexed some muscles and conducted sea trials with the Liaoning aircraft carrier. Finally, Beijing announced the creation of two new administrative districts in the South China Sea, covering disputed maritime features.[24] The choppiness of SCS, mentioned earlier in this book, is growing continuously and IOR has to be prepared to face some tempestuousness.

The violations are not limited to the maritime domain. China's self-declared East China Sea ADIZ, covering a large swath of airspace over the East China Sea (including over small islands that are the subject of a territorial dispute among Japan, the PRC, and Taiwan) has become more turbulent. Violations in the conventional ADIZs are on increase. China is now ratcheting up the risk of military confrontation in the Taiwan Strait, as Beijing seeks to deter Taipei from continuing to deepen ties with the US and other like-minded democracies. PLAAF aircraft have repeatedly breached the median line between Taiwan and the Chinese mainland in month of September 2020, in the garb of a series of military exercises in the area. The aircraft included fighter jets and bombers[25] which had put AD weapons of Taiwan on high alert. Japanese and Vietnamese air forces also have been witnessing increased number of air space violations by PLAAF aircraft especially over the contested island territories in the SCS. China is dealing harshly with Hong Kong as well and has been criticised globally. The new national security bill implemented by the Chinese Communist Party (CCP) ensures the semi-autonomous state is absorbed completely within the Chinese jurisdiction. The legislation will effectively strip residents of Hong Kong of their freedom and rights.[26]

Besides the wolf warrior diplomacy and military coercion strategy, the 'debt-trap' policy of China has been clearly entrapping countries into shackles of heavy loans specially those who have supported BRI. BRI also has come under fund crunch as China is paying attention on domestic revival plans. Hence the countries that are part of this debt trap are expected to fund the projects which is clearly not their cup of tea. Project failures, cases of insolvency, and bankruptcies are expected to grow exponentially along BRI routes in the months, if not years, to come.[27] It is becoming a catch 22 situation for China and the anxiety is evident in form of aggressiveness on all fronts.

As a result, Beijing's unconcealed belligerence is prompting unceasing debates about the underlying costs of reliance on China and spurring support for closer coordination between other Indo-Pacific partners. The Indian, Japanese, Malaysian, and Australian governments have already taken concrete steps to reduce their economic exposure to Beijing, spanning investment, manufacturing, and technology.[28] China's bullying tactics and duplicity in the South China Sea, it's stealing of technologies, refusal to open its markets and aggressive diplomacy has aggravated many countries, which had earlier looked upon China's rise rather benignly. These would include the UK, Canada and even the EU. Countries are taking steps to decouple themselves from China economically. The on-going US-China trade dispute had already prompted companies to re-think their investments in China.[29]

Effects on the IOR

There is a growing deliberation on the future of post-COVID world order in terms of geopolitical realignments and the emergence of new partnerships across the world. The IOR is also not unaffected and the pandemic is certainly going to further accelerate the shift in the regional balance of power within the IOR. Based on national interests and respective needs of the countries, they are trying to realign and do the balancing with major powers in the IOR so as to maximize their respective gains. The pandemic has impacted the economies of all countries in the region and this downturn may be long lasting for the years to come. It is possible that many of interventions done by the leadership of individual country might not produce desired results and this situation will ultimately result in slower growth in near future. This void may be utilised by the major powers to compete for geopolitical influence in the countries in the IOR and it also raises the chances of sovereign debt crises.[30]

To bolster its image in the IOR, China focused on another strategy, the so-called 'mask diplomacy' by donating relatively small amounts of testing kits and protective clothing to several countries including Pakistan, Bangladesh, Nepal, Sri Lanka, Mauritius and Maldives. China also made commercial sales of much larger quantities of supplies, some of which were of questionable quality. China is also indicating some willingness in principle to provide debt relief to low income countries in the IOR, but it is to be seen how much relief it would be willing to give—and whether it would seek strategic quid pro quos from borrowers or not, like it has been doing under its debt-trap policy. Despite significant disruptions to many projects, Pakistan, China's principal strategic partner in the region, is pressing ahead with the CPEC. If anything, Pakistan is likely to double down on CPEC, reflecting how much successive governments have gambled on it as the country's saviour. At the same time, Pakistan has already asked China for debt rescheduling on USD 30 billion in projects, which it is likely to obtain given the political importance of CPEC to Beijing.[31] In fresh inputs now available in open source domains, China and Iran are on course to chalk out a comprehensive 25 year strategic partnership to infuse USD 400 billion worth of investments in Iran's oil and gas, infrastructure and transportation sectors. China will position up to 5000 Chinese security personnel to protect its projects in Iran and military cooperation is also likely to increase as per reports.[32] China's access to the IOR via Gwadar in Pakistan, its military base at Djibouti and possible permanent military access arrangement with Iran will have lasting implications on stability of the region.[33] On the other hand, there are few positive signs of enhanced stability in the Western IOR because of the US brokered peace between Israel and two Arab countries—the United Arab Emirates and Bahrain (these three are IOR countries). This accord is revolutionary in scope as it aims at normalising ties and focusing on business, trade, and travel. These deals not only set the stage for a sea change in Arab-Israeli relations, but also have the potential to present a novel opportunity to finally solve the most intractable issue between the two sides: the Israeli-Palestinian conflict.[34] However, this sudden phase of harmony has been a cause of headache for many countries including Turkey and Iran which are potential destabilisers of the region. The fact that the distance to Iran's nuclear sites is lesser from UAE than Israel along with the existing US sanctions, are making Iran restless. Pakistan, after failing to register any objection from any of its key allies within OIC (Organisation of Islamic Countries) regarding the Kashmir issue (Article 370) has sought company of Turkey and Malaysia. China is raising issues regarding Indian sovereignty over the Union Territory of Ladakh and hence supporting Pakistan by all means. Therefore, a nefarious nexus of Turkey, Iran, Pakistan, Malaysia (with Chinese overtures) is a credible threat for stability of the IOR.

India's Regional Response to COVID-19

India's assistance to other countries, especially in the IOR, amid the pandemic has been appreciated world over. India's approach, including on the diplomatic front, has been proactive since the outbreak of the crisis. One of the first steps taken by India was to evacuate citizens of different countries along with its own citizens from Wuhan, the epicentre of the COVID-19 outbreak. Those evacuated by Air India and the IAF as compassionate cases included citizens from IOR countries such as Bangladesh, Madagascar, Maldives, Myanmar, South Africa etc. India not only evacuated these people, but also quarantined them in India (even at the IAF bases) as a precautionary measure before sending them to their respective countries.

India has also emerged as a major supplier of medicines to different countries worldwide in the fight against COVID-19. As part of this effort, India was the first responder to Mauritius and Seychelles. India has assured all possible help to both these countries to face the challenge of COVID-19. India also sent the IAF aircraft to Maldives and Nepal with medical help. COVID-19 has provided India an opportunity to address the irritants in its ties with some countries, most notably Iran and Malaysia. India and Iran cooperated with each other in order to evacuate the Indians stranded in Iran. Later Iranian President wrote to the Indian PM seeking India's assistance to deal with COVID-19. Apart from this, India has sent a wheat consignment to Afghanistan through Iran's Chabahar port. With respect to Malaysia, India has agreed to supply anti-malarial drugs, indicating improvement in the bilateral relations. In all, India in month of May 2020, had decided to extend its medical assistance effort to help other countries in combating COVID-19 and had targeted covering more than 90 countries. This diplomatic outreach was expected to cost Rs 110-120 crore (Rs 1.1-1.2 billion).[35] As the biggest financial help to a country during pandemic, India handed over financial assistance of USD 250 million to the government of Maldives in September as budgetary support to mitigate the economic impact of the COVID-19 pandemic.[36] With the growing requests from all over the world, India sent medical supplies to over 150 countries during the coronavirus crisis. The Prime Minister in his virtual address to the United Nations General Assembly (UNGA) said that India can help in bringing the world out of the coronavirus crisis with mass vaccine distribution once all trials are completed successfully.[37] India also handed over medical supplies to the Government of Myanmar which included around 200,000 tablets of Hydroxychloroquine (HCQ), gloves, body bags, syringes, defibrillators and more to give a boost to Myanmar's efforts to fight the pandemic.[38] India and Myanmar also have decided to collaborate on development of COVID-19 vaccine.

India's outreach in the IOR and the broader Indo-Pacific Region has been parallel to China's outreach by way of the BRI. India's competition with China could be considered asymmetric owing to differences in the economic and military capabilities of both countries. However, China's development projects have resulted in indebtedness for many countries with limited resources. This situation creates an opportunity for India to present its own alternative to countries in the Indian Ocean Region, which would also facilitate increasing India's footprint in the region. India's diplomacy in the IOR could define its position in the changed world order.[39] Like India, other countries within the IOR will have to find a way out to fight such economically crippling crises because unless they change path and follow a totally different economic model based on self-reliance, the gap between slow-growing countries as compared to economically dynamic countries elsewhere in the world is likely to widen.[40]

While focussing on diplomacy through SAGAR in post-COVID IOR, India should also strengthen its military diplomacy. In April, the Indian Navy issued a public statement saying that its Eastern Naval Command's Dornier squadron continues its maritime surveillance missions and that its naval assets are always ready to be deployed. The statement was likely prompted by the discovery of a dozen China-deployed 'Sea Wing' underwater drones in the eastern Indian Ocean, India's traditional sphere of influence. The drones are Unmanned Underwater Vehicles (UUVs) and are purportedly used for scientific research for deep-sea mining and other commercial activities. However, they can also be used to facilitate submarine movements and, in case of a conflict, discover and terminate underwater mines. This deployment of research vessels, UUVs and submarines in the Indian Ocean Region, even when the COVID-19 virus was devastating the country, did not show any decline. The Indian Navy may still have numerical superiority in terms of vessels and manpower, but China, with or without the displacement and disruption caused by the virus, is laying the groundwork to rapidly expand its presence in the region. One should not be surprised if China uses the global confusion (caused by the COVID-19 crisis) to its advantage and expands its sphere of influence in the Indian Ocean Region under the garb of relief work. Aid politics has become a new phenomenon in the Indian Ocean amid the COVID-19 crisis.[41]

Response by India on Domestic Front

Various ministries and government organisations in India have worked together under effective leadership and have prevented a larger adverse effect of this pandemic so far. Because of this multi-dimensional approach, India has been able to fight Corona not only domestically but also assisted other nations which needed some succour. Along with Ministry of Home Affairs, Health and External Affairs, the Indian Armed Forces have contributed significantly in this fight against Corona. The COVID-19 pandemic has been used as an opportunity to work towards making India 'self-reliant' while providing all possible help to its regional neighbours by extending medical and other assistance to many countries in the fight against the COVID-19 pandemic.[42] To fight with the pandemic, domestic production of masks, gloves and Personal Protective Equipment (PPE) suits has gone up enormously and India became second largest PPE manufacturer of the world in just 60 days (Fig. 9.3).[43] It is because of these measures and systematic lock down plan that as against the death rate of 350 individuals per million in the US and over 600 per million in European nations like the UK, Italy and Spain, the rate of fatalities in India was less than 12 per million.[44]

Fig. 9.3: PPE Production in India per day
From 0, India now produces nearly 4.5 lakh PPE kits every single day.

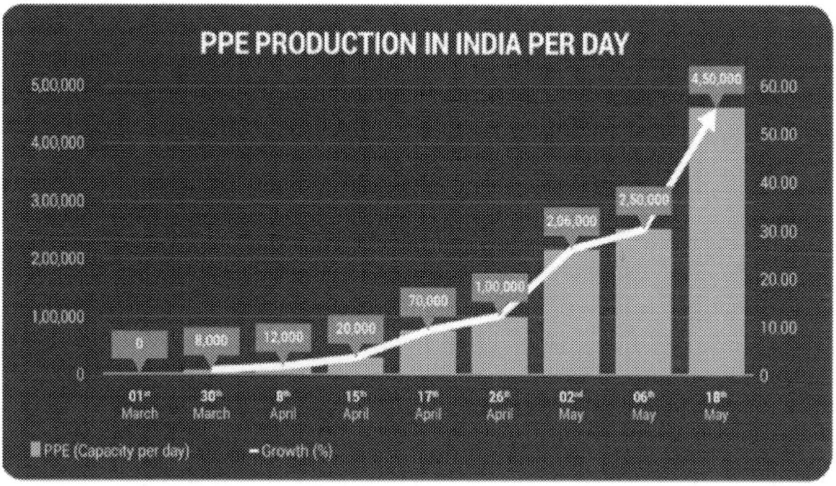

Source Image: Strategic Investment Research Unit

India's performance in combating the health crisis has been significant and certainly the loss of lives has been controlled effectively.[45] The graph of fatality rate is an indication of this performance (Fig. 9.4).

Fig. 9.4: Fatality Rate: Amongst the Lowest in the World

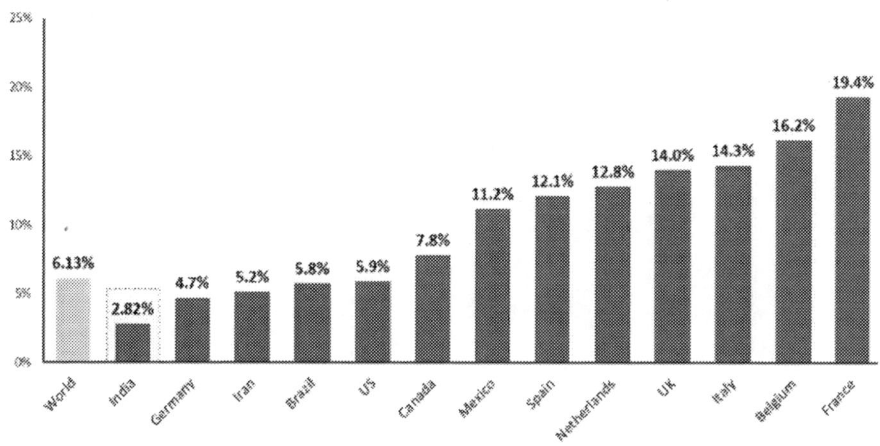

Source: https://zeenews.india.com/india/india-s-covid-19-fatality-rate-at-2-82-senior-citizens-constitute-50-of-deaths-2287542.html

Role of the IAF in Fight against Corona

This being another catastrophic situation for the country, the IAF, along with the Army and the Navy, rose to the occasion and has been providing the required 'speed' and 'mobility' to efforts of the government in country's fight against COVID-19. As the crisis began to unfold in the country, the IAF put all its 'air mobility' assets at the disposal of the nation's critical need. As early as in February 2020, the IAF had swung in action to help the government in tackling the pandemic. On 26 February 2020, 76 Indians and 36 foreign nationals were evacuated from Wuhan (China) in a C-17 aircraft. The other people were from Bangladesh, Madagascar, Maldives, Myanmar, South Africa and the US. The aircraft had also carried medical supplies as symbol of goodwill to China. The IAF brought back 58 Indians from Iran on 10 March 2020, who were stranded there due to the COVID-19 outbreak. The IAF, as part of the defence services' support to civil authorities, has been continuing with unabated relief measures. Operation 'Sanjeevani' was conducted effectively with efficient deployment of a large number of wide-bodied aircraft and helicopters all across the country to shift medical supplies and associated equipment. Under this operation, the IAF sent aircraft to

Fig. 9.5: An IAF C-17 Evacuating Indian Citizens from Iran

Source: https://www.businesstoday.in/latest/trends/coronavirus-update-india-evacuates-first-batch-of-58-indian-citizens-from-iran/story/398007.html

Maldives and Nepal with medical help and is likely to be tasked for more such missions. Since 24 March 2020, the IAF has flown approximately 400 sorties ferrying over 700 tonnes of medicine, ration and other essential supplies to various parts of the country.[46] Towards this effort, approximately 1300 passengers and 10000 test swabs were also transported by the IAF. In this war, the 'air warriors' along with the valiant healthcare professionals have been at the forefront extending continuous support to the nation, immaterial of the dangers they face themselves from this pandemic.[47]

China Provokes India: Aggression along the LAC during Pandemic

India hasn't been unaffected by the reckless post-COVID behaviour of China. In an unprecedented demonstration of aggressiveness that too in difficult time of global pandemic, PLA carried out multiple incursions leading to standoffs at Pangong Tso, Hot springs and Galwan in Eastern Ladakh and Naku La in Sikkim. These incursions, unlike the previous ones at Depsang (2013), Chumar (2014), and Doklam (2017) were greater in intensity, indicating long haul of China. The on-going border dispute turned hostile after Galwan clash on 15 June 2020 which was the bloodiest clash after 1967. The PLA troops laid a deliberate ambush, breaching the protocols, attacking the patrol party that had gone to check on the agreed-upon disengagement in Galwan. PLA troops assaulted Indian soldiers with the most primitive weapons like spiked clubs killing 20 soldiers. Without resorting to use of conventional weapons, the Indian troops gave a befitting reply to the PLA reportedly causing some 40 odd fatalities.[48] The Army further bolstered its dominance over a number of strategic heights overlooking key Chinese-held positions around the Pangong lake area in eastern Ladakh when China unsuccessfully attempted to occupy Indian territory in the southern bank of Pangong lake on the intervening night of 29 and 30 August.[49]

China is applying its military coercion approach in Indian context. Unlike 1962, where Indian side was relying on diplomatic talks alone and was not prepared for assault, the Indian Armed Forces were able to counter Galwan aggression and subsequently were put on high alert immediately. The IAF was quick to deploy and start patrolling the skies of eastern Ladakh. Transport aircraft and helicopters of the IAF assisted the Army in troop build-up at a rapid pace along with supplying necessary material. The selected fighter squadrons as well as helicopter units were deployed in no time. Newly inducted Chinook helicopters were seen lifting the army machinery while Apaches were also deployed at nearby airfields. Air Chief Marshal RKS Bhadauria gave a statement about complete readiness of the IAF, "It should be very clear that we are well prepared and suitably deployed to respond to any contingency. I assure the nation that we are determined to deliver and will never let the sacrifice of our braves of Galwan go in vain," he said.[50]

The most probable reason for this aggression is that China has invested heavily in BRI and hence is trying to coerce India into accepting the BRI or at best not to oppose it. CPEC, which is central to the BRI, is China's gateway to Eurasia via Central Asia. It is strategically very important for the CCP's dream of seeing China as the largest economy of the world. India's demonstration of political will by abrogating the provisions of Article 370 on 05 August 2019, carving out two Union Territories of J&K and Ladakh from the erstwhile state, and rightly seeking full control of its occupied territories[51] has caused unrest in headquarters of the CCP and hence the knee jerk reaction at LAC has increased the tension. In the post-COVID world order wherein India is likely to emerge stronger and a better destination for investment, China can do nothing but create some sense of insecurity.

Resilient Time Ahead: What Should India Do?

It is not merely a military face-off problem. The muscle flexing is only a part of larger strategy which emanates from the CCP headquarters and consequent power demonstration by India can be sign of determination but not the complete solution. The political resolution of disputed borders along the LAC has to be achieved as permanent solution to this cancerous problem which keeps relapsing. The first step has to be the restoration of the status quo ante that prevailed before the present crisis and implementation of the 1993 agreement regarding the identification of LAC. India must call out China's deceit and actively engage other nations that view China as a major threat to stability and peace.[52] Quadrilateral Security Dialogue (Quad) is an option to strategically contain China and India can play a major role in the Indo Pacific. By inviting Australia in Malabar exercise 2020, India has firmly augmented Quad's strategic importance in the backyard of China. The decision to add Australia made the iteration of Malabar the first exercise to include all four, or "Quad," states since the grouping's reconvening in November 2017 after a decade-long hiatus.[53] Quad members should actively cooperate to build each other's economic, technological and military capabilities to counter China. Without this enhancement, the Quad's goals—maintaining Indo-Pacific stability, challenging regional bullying, strengthening economic independence, and protecting the free and open Indo-Pacific under the rules-based order—will remain a distant dream.[54] With Australia, Japan and the US already limiting China's manoeuvres, India must stand stern diplomatically and deter militarily as a responsible Asian democracy of the IOR. For this India shall take few steps mentioned below.

Gain Adroitness in Space

India needs to openly build and demonstrate its strength, especially military prowess and intent to use it not as the last resort but as an essential one to be used from the beginning, along with other instruments of power, especially when territorial integrity is threatened. Preparedness has to become the primary means of deterrence. Preparedness, not only in terms of infrastructure (like roads, tunnels and airfields) but also in terms of its synonym, vigilance, will have to be enhanced. India needs to enhance the ISR capabilities for effective management of all borders including maritime AOR like the Indian Ocean. In such vast spaces, occupied by the steady passages of merchant shipping through the SLOCs, the requirement for impeccable intelligence inputs cannot be overemphasised. Satellite coverage, to oversee the large tracts of ocean, becomes an absolute necessity.[55] Satellites have been proven very effective for task of ISR. With the combination of Electro-Optic, Synthetic Aperture Radar and Electronic Intelligence sensors, these satellites can gather phenomenal intelligence useful for all purposes. China has reportedly launched Yaogan 30A, 30B, 30C satellites in 2017 and their trajectory has been confirmed over Ladakh region few days before the dastardly attack.[56] The latest launch of military satellites by China was on 24 March 2020. Long March 2C rocket was used to carry sixth group of three Yaogan-30 ISR satellites from Xichang in Sichuan province.[57] India must strengthen its Defence Space Agency with full support of ISRO and work towards gaining self-reliance in space domain. The IAF has the capability to lead the transformation from space agency to space command in future as the future lies in the command of aerospace. Many countries have dedicated space commands under their air forces to handle this crucial task which are the key agencies responsible for strategic, operational and tactical intelligence.

Pace up Indigenisation

India should also focus on maximum indigenisation of defence manufacturing sector. This economically strong largest democracy must be independent in this sector so as to avoid relying on other countries during crises. India is the largest importer of arms with nearly 60 per cent of the military hardware of Russian origin and diversification is achieved from the US, Israel and France. The lack of a defence industrial base also forces the Armed Forces to store an unnecessarily huge inventory as War Wastage Reserves.[58]

Prepare for Hybrid Warfare as a Nation

India, conclusively, should prepare to mitigate threats emanating from Hybrid Warfare which almost every country is facing. The present situation is being called a 'No War No Peace' situation. The conventional wars are only subset of hybrid war-the term which attempts to blend conventional and irregular warfare, at times superimposed by belligerent activities in the cyber domain. Banning of Chinese applications to prevent theft of data is a small step but in correct direction because data gives the adversaries information which can be used to prepare weapons of Information Warfare (IW), another component of hybrid war. In today's time IW can also be called war of narratives in which a different image is projected in rest of the world so as to create a bias against the target country e.g. situation in Kashmir for many countries is equal to situation in war torn states of the world because of the narrative created by anti-India nations/elements.

There is a need of structures and systems at the national level as also within the armed forces to counter IW. Cyberspace, Artificial Intelligence, advanced computing, mobile networks, unmanned and autonomous systems, and social media present a military revolution in this kind of warfare. To extract its full potential, militaries need to adapt institutional orientation towards non-lethal warfare. This can be achieved through better training and education, and deliberate integration of information operations across the military services during planning and operations. The nation and armed forces need IW structures to effectively exploit this domain as an integral component of its war fighting strategy to counter the inimical designs of the adversaries.[59] The Defence Cyber Agency has a dedicated domain of cyber space to look into however the broader domain of hybrid warfare of which, IW is a significant part, needs special combination of experts ranging from domain experts in uniform, from think tanks and from distinguished media group in order to counter the adversaries in invisible battlefield.

There is no doubt that the Indian Armed Forces are combat rich and battle hardened, as also the best force for high altitude warfare. The IAF has 'quality over quantity' advantage over PLAAF in the Tibetan plateau region. The existing fighter fleet with newly inducted Rafale aircraft maintains capability to deter Chinese air aggression and advanced helicopters like Apache and Chinook along with Mi-17 V5 has credibility to support all kinds of battlefield operations of highly formidable Indian Army. Similarly, the IAF and the Navy have the capacity to jointly protect maritime borders and preserve stability in the IOR with other defence forces. However, the armies, air forces and navies do not fight each other on one-on-one wars, nations go to war. China's non-contact warfare (hybrid warfare) capabilities need to be assessed and own defensive offensive capabilities need to be enhanced to mitigate the threat with a consistent joint planning.[60]

Conclusion

Even as the initial impact of COVID-19 appears to be fading out, it is quite evident that coronavirus is not going away for a while. As citizens adjust to the new normal, India's international relations also need to adjust to the same. It is unclear how overall global relations will be post-COVID-19. This means that India must take time now to build new as well as redefine old international relations.[61] China's unusual aggression is a litmus test of India's diplomacy as well as defence preparedness. India's tough stance can prove to be inspiration for other countries subject to intimidation. The SAGAR doctrine must be used pervasively because the bilateral and multilateral relationships will only provide stability in the region and will allow generation of collective deterrence against nefarious designs. A concord of democracies will be required to create outstanding strategic equilibrium in the IOR and Indo Pacific to ensure sustained stability. India, with its political will, diplomatic relations, economic resources, and military resilience, can become a crucial player of such a concord. It is time when India rises to the occasion.

NOTES

1. Amin Bagheri, https://www.newsgram.com/coronavirus-bioweapon/ (Accessed 15 June 2020).
2. "WHO declares Covid-19 outbreak a pandemic", 12 March 2020, https://www.pharmaceutical-technology.com/news/who-declares-covid-19-pandemic/ (Accessed 15 June 2020).
3. Sudhanshu Tripathi, https://timesofindia.indiatimes.com/blogs/abhiram/is-corona-virus-a-new-bio-weapon/ (Accessed 20 June 2020)
4. Covid-19 Coronavirus Pandemic, https://www.worldometers.info/coronavirus/ (Accessed 30 October 2020).
5. Ibid.
6. Corona Tracker: India Overview, https://www.coronatracker.com/country/india/ (Accessed 30 October 2020).
7. Tripathi, op. cit.
8. Lora Jones, Palumbo Daniele and Brown David, "Coronavirus: A visual guide to the economic impact", *BBC News*, https://www.bbc.com/news/business-51706225#:~:text=Risk%20of%20recession&text=But%20the% 20IMF%20says%20 thatG reat%20Depression%20of%20the%201930s.&text=Recovery%20in%20big% 2C%20 services%2 Dreliant,to%20be%20a%20slow%20process. (Accessed 01 August 2020).
9. Explained desk, "Explained:how has covid-19 affected the global economy", https://indianexpress.com/article/explained/explained-how-has-covid-19-affected-the-global-economy-6410494/ (Accessed 10 August 2020).
10. World Economic Forum report, "COVID-19 hit commodities hard - these 8 charts show what we can expect", https://www.weforum.org/agenda/2020/11/commodity-markets-economics-charts-oil-energy-metals-gold-food-covid-19-coronavirus-pandemic/ (Accessed 02 November 2020).
11. Katharina Buchholz, "Coronavirus Slump Continues For GDP Growth", https://www.statista.com/chart/18095/quarterly-gdp-growth-predicted-growth-selected-industrialized-nations-oecd/ (Accessed 11 November 2020).
12. "India Real GDP Growth, 2005–2020", *CEIC Data*, https://www.ceicdata.com/en/indicator/india/real-gdp-growth (Accessed 15 October 2020).

13. *The Economist*, "Would-be autocrats are using covid-19 as an excuse to grab more power", https://www.economist.com/international/2020/04/23/would-be-autocrats-are-using-covid-19-as-an-excuse-to-grab-more-power (Accessed 08 July 2020).
14. "Resolution 2532 (2020) Adopted by the Security Council on 1 July 2020", https://undocs.org/en/S/RES/2532(2020) (Accessed 10 July 2020).
15. Katariina Mustasilta, "From Bad to Worse?: The Impact(s) of Covid-19 on Conflict Dynamics", *European Union Institute for Security Studies (EUISS)*, 2020, www.jstor.org/stable/resrep25022 (Accessed 10 July 2020).
16. Kapur Lalit, "Geopolitical Lessons of COVID-19 for India", DPG policy papers, https://www.delhipolicygroup.org/publication/policy-papers/geopolitical-lessons-of-covid-19-for-india.html (Accessed 10 May 2020).
17. Li Yuan, "Coronavirus Crisis Shows China's Governance Failure", https://www.nytimes.com/2020/02/04/business/china-coronavirus-government.html?auth=login-google (Accessed 22 August 2020).
18. David Brennan, "China's Mix of Aggressive and Charming Coronavirus Diplomacy could Worsen Europe's Fractures, Report warns", https://www.ifri.org/en/espace-media/lifri-medias/chinas-mix-aggressive-and-charming-coronavirus-diplomacy-could-worsen (Accessed 10 May 2020).
19. Kapur, op. cit.
20. Tripathi, op. cit.
21. Lindsey Ford and Gewirtz Julian, "China's post-COVID aggression is reshaping Asia", https://www.thehindu.com/opinion/lead/chinas-post-covid-aggression-is-reshaping-asia/article32095103.ece (Accessed 05 August 2020).
22. Shakti Sinha, "China: Anxiety and Aggression", in *One Mountain Two Tigers* (New Delhi: Pentagon Press LLP, , 2020), p. 182.
23. Geeta Mohan, "China's growing territorial aggression amid coronavirus crisis", https://www.indiatoday.in/mail-today/story/china-growing-territorial-aggression-amid-covid-crisis-1692787-2020-06-26 (Accessed 09 July 2020).
24. Zachary Williams, "China's Tightening Grasp in the South China Sea: A First-Hand Look'" *The Diplomat*, https://thediplomat.com/2020/06/chinastightening-grasp-in-the-south-china-sea-a-first-hand-look/, (Accessed 10 July 2020).
25. *Bloomberg News*, "Chinese aircraft breach Taiwan airspace; raise risk of clash in the region", https://www.business-standard.com/article/international/chinese-aircrafts-breach-taiwan-airspace-raise-risk-of-clash-in-the-region-120092100320_1.html (Accessed 15 October 2020).
26. Mohan, op. cit.
27. Plamen Tonchev, "The Belt and Road After COVID-19", https://thediplomat.com/2020/04/the-belt-and-road-after-covid-19/ (Accessed 11 June 2020).
28. Lindsey Ford and Gewirtz Julian, op. cit.
29. Sinha, op. cit., p. 191
30. Siddharth Singh and Deeksha Goyal, "Report: Impact of COVID-19 on the countries of Indian Ocean Region", https://indiafoundation.in/wp-content/uploads/2020/06/Impact-of-COVID-19-on-the-countries-of-Indian-Ocean-Region.pdf (Accessed 01 August 2020).

31. David Brewster, "Covid contest in Indian Ocean region: India, China jostle for top spot", *The Interpreter*, https://www.lowyinstitute.org/the-interpreter/covid-contest-indian-ocean-region-india-china-jostle-top-spot (Accessed 20 June 2020).
32. NK Bhatia, "Iran-China strategic agreement and its implications for India", https://www.financialexpress.com/defence/iran-china-strategic-agreement-and-its-implications-for-india/2022453/ accessed 20 July 2020)
33. Kabir Taneja, "Reading a potential Iran–China strategic agreement from India", https://www.orfonline.org/expert-speak/reading-a-potential-iran-china-strategic-agreement-from-india/ (Accessed 20 July 2020).
34. Koduvayur Varsha and Daoud David, "Welcome to a Brand-New Middle East", https://foreignpolicy.com/2020/09/30/israel-uae-bahrain-palestinians-peace/ (Accessed 11 October 2020).
35. Shishir Gupta, "India draws up Rs 1 billion Covid-19 medical assistance plan, targets 90 countries", https://www.hindustantimes.com/india-news/india-amps-up-covid-19-medical-assistance-plan-targets-to-reach-90-countries/story-0X1H8z1Zqi9piw6n4FDu8J.html (Accessed 30 September 2020)
36. Rekha Dixit, "India provides $250 million aid to Maldives to mitigate COVID-19 impact", https://www.theweek.in/news/india/2020/09/20/india-provides-250-million-aid-to-maldives-to-mitigate-covid-19-impact.html (Accessed 30 September 2020).
37. Debanish Achom, https://www.ndtv.com/india-news/as-biggest-vaccine-maker-india-will-help-world-overcome-covid-pm-narendra-modi-to-un-general-assembly-2301496 (Accessed 30 September 2020)
38. Huma Siddiqui , "Coronavirus Pandemic: India lends helping hand to it's neighbour Myanmar; send medical supplies", https://www.financialexpress.com/defence/coronavirus-pandemic-india-lends-helping-hand-to-its-neighbour-myanmar-send-medical-supplies/1948830/ (Accessed 30 September 2020)
39. Niranjan Marjani, "India's Indian Ocean Diplomacy in the COVID-19 Crisis", *The Diplomat*, https://thediplomat.com/2020/04/indias-indian-ocean-diplomacy-in-the-covid-19-crisis/ (Accessed 30 June 2020).
40. Goyal Singh, loc. cit.
41. Bertil Lintner, "China eyes a Covid-19 edge in the Indian Ocean", *Asia Times*, https://asiatimes.com/2020/04/china-eyes-a-covid-19-edge-in-the-indian-ocean/ (Accessed 25 July 2020).
42. ANI, "India has extended assistance to over 150 countries in fight against COVID-19: PM Modi", https://health.economictimes.indiatimes.com/news/policy/india-has-extended-assistance-to-over-150-countries-in-fight-against-covid-19-pm-modi/77028422 (Accessed 18 September 2020).
43. Remya Lakshmanan and Mishika Nayyar, "Personal Protective Equipment in India: An INR 7,000 Cr industry in the making", https://www.investindia.gov.in/siru/personal-protective-equipment-india-INR-7000-cr-industry-in-the-making (Accessed 20 September 2020).
44. PTI, "India's fight against Covid-19 driven by its people: PM Modi", https://theprint.in/india/indias-fight-against-covid-19-driven-by-its-people-pm-modi/450183/ (Accessed 20 July 2020).

45. Raghbendra Jha, "India's Response to the Coronavirus Crisis", *Australian Outlook, AIIA*, http://www.internationalaffairs.org.au/australianoutlook/indias-response-to-the-coronavirus-crisis/ (Accessed 21 September 2020).
46. Vikas Kalyani, "The Indian Air Force and an increasing HADR mandate", https://www.orfonline.org/expert-speak/indian-air-force-increasing-hadr-mandate-65992/ (Accessed 15 May 2020).
47. Sameer Pradhan, "Fighting COVID-19: IAF'S response to the Coronavirus threat", https://www.financialexpress.com/defence/fighting-covid-19-iafs-response-to-the-coronavirus-threat/1991320/ (Accessed 20 June 2020).
48. Bhatia Vinod, "Post Galwan: Deter China's Aggressive Behaviour", in Shakti Sinha (Ed), *One Mountain Two Tigers* (New Delhi: Pentagon Press LLP, 2020), p. 106
49. PTI, "Indian Army further strengthens dominance in strategic heights in eastern Ladakh", https://www.deccanherald.com/national/indian-army-further-strengthens-dominance-in-strategic-heights-in-eastern-ladakh-885596.html (Accessed 20 September 2020).
50. Srinivasa Rao Apparasu, *Hindustan Times*, https://www.hindustantimes.com/india-news/not-at-war-with-china-but-fully-prepared-says-air-force-chief-bhadauria/story-Lhs3mpn1DfBIGcWkmMTUnM.html (Accessed 20 June 2020).
51. Bhatia, op. cit., p. 109
52. Ibid
53. Ankit Panda, "Australia Returns to the Malabar Exercise", https://thediplomat.com/2020/10/australia-returns-to-the-malabar-exercise/ (Accessed 30 October 2020).
54. Harsh V Pant and Mann Anant Singh, "India's Malabar Dilemma", ORF Issue Brief issue no. 393, August 20, https://www.orfonline.org/wp-content/uploads/2020/08/ORF_IssueBrief_393_Malabar.pdf (Accessed 20 August 2020).
55. Sumit Mukerji, "Dominating the High Ground Across the Indian Ocean Region", https://salute.co.in/dominating-the-high-ground-across-the-indian-ocean-region/ (Accessed 30 June 2020).
56. Rishab Gulati, "The truth behind China's Ladakh Spy sats and how it planned Galwan heist", https://www.newsx.com/national/the-truth-behind-chinas-ladakh-spy-sats-and-how-it-planned-galwan-heist.html (Accessed 28 July 2020).
57. "China's Long March 2C Rocket Launches Military Surveillance Satellites Into Orbit", https://spacewatch.global/2020/03/chinas-long-march-2c-rocket-launches-military-surveillance-satellites-into-orbit/ (Accessed 30 July 2020).
58. Bhatia, op. cit., p. 116.
59. Ibid, p. 117.
60. Ibid.
61. Priyanka Deo, "Steps India must take for strong foreign relations post COVID-19", https://www.wionews.com/opinions-blogs/steps-india-must-take-for-strong-foreign-relations-post-covid-19-304818 (Accessed 15 July 2020).

Appendix 1

TEXT OF THE PM'S SPEECH IN MAURITIUS INTRODUCING VISION 'SAGAR'

Hon'ble Prime Minister Sir Aneerood Jugnauth, distinguished guests,

To our naval personnel from India, our guardian of the seas, who are here today – my special greetings to you.

It is a great honour to commission the Barracuda today in the service of the National Coast Guard of Mauritius.

Thank you for giving me this privilege. Thank you for choosing India as your partner.

This ship has made a long journey – from Kolkata across the great arc of Indian Ocean to this beautiful shore.

Generations ago, people from India charted that course to a new destination and a new life.

Today, Barracuda brings with her the goodwill and greetings of the people of India.

She represents our unique trust and confidence.

She reflects our partnership that has stood the test of time.

She is a symbol of our shared commitment to peace and security in Indian Ocean- our common maritime home.

Barracuda is a beautiful ship. She is also very capable; and built to specifications of Mauritius.

Now, she sails proudly with the flag of Mauritius. She will protect your islands and your waters. She will be there to help in times of distress and emergencies.

But, she will do more than that. She will also help make our Indian Ocean safer and more secure.

In doing so, Mauritius will fulfill an important international responsibility because, the Indian Ocean is critical to the future of the world. This Ocean bears two-thirds of the world's oil shipments, one-third of its bulk cargo; and half of its container traffic. Over three-fourths of its traffic goes to other regions of the world.

The vast Indian Ocean Region hosts over 40 states and nearly 40% of the world's population. It touches Australia, Southeast Asia, South Asia, West Asia and the eastern seaboard of Africa. It sparkles with the gems of island states.

Think of the civilisational links, yet great diversity in this vast region! Consider the vast opportunities that it holds!

Today, the world speaks of 21st century driven by the dynamism and the energy of Asia and the Pacific. But, its course will be determined by the tides of the Indian Ocean. This is why Indian Ocean is at the centre of global attention more than ever before.

We also see growing global stakes and presence in the Ocean. Even in this changing world, the oceans hold the key to its fortunes. And, we will all prosper when the seas are safe, secure and free for all.

To ensure this will be our greatest collective responsibility. But, we will also have to rise to other challenges that are not uncommon in our region.

We have seen the tragedy of tsumanis and cyclones.

Terror has visited us from sea. Piracy has taken toll of lives and trade on a scale that seemed beyond belief in modern era.

Illegal fishing and oil spills hurt our economies. We all feel the rising impact of climate change on our coasts and islands.

We have also seen that instability and turmoil in coastal and island states can have profound impact on security on the seas.

India is at the crossroads of Indian Ocean.

Since Lothal in Gujarat became one of the earliest seaports in the world, India has had a long maritime tradition.

Our cultural footprints stretch across Asia and Africa. We see this in our strong Diaspora across oceans.

The seas forged links of commerce, culture, and religion with our extended neighbourhood across several millenniums.

Our more recent history has focused our attention on our continental neighborhood.

But, India has been shaped in more ways by the seas around us.

Today, 90% of our trade by volume and 90% of our oil imports take place through sea. We have coastline of 7500 km, 1200 islands and 2.4 million square kilometers of Exclusive Economic Zone.

India is becoming more integrated globally. We will be more dependent than before on the ocean and the surrounding regions.

We must also assume our responsibility to shape its future.

So, Indian Ocean Region is at the top of our policy priorities.

Our vision for Indian Ocean Region is rooted in advancing cooperation in our region; and, to use our capabilities for the benefit of all in our common maritime home.

It means many things.

One, we will do everything to safeguard our mainland and islands and defend our interests.

Equally we will work to ensure a safe, secure and stable Indian Ocean Region that delivers us all to the shores of prosperity.

And, our capabilities will be there for those struck by the ocean's fury. Or, caught in distress on the seas.

Second, we will deepen our economic and security cooperation with our friends in the region especially our maritime neighbours and island states.

We will also continue to build their maritime security capacities and their economic strength.

Three, collective action and cooperation will best advance peace and security in our maritime region. It will also prepare us better to respond to emergencies.

That is why, in 2008, India promoted the Indian Ocean Naval Symposium. Today, it brings together 35 navies of the region. Our goal is to deepen our mutual understanding on maritime challenges; and, strengthen our collective ability to address them.

We also support efforts to strengthen our regional mechanisms for maritime cooperation – from dealing with piracy terrorism and other crimes; to marine safety and natural disasters.

India has also started maritime security cooperation with Maldives and Sri Lanka and we hope that Mauritius, Seychelles and other nations in the region will also join this initiative.

Four, we also seek a more integrated and cooperative future in the region that enhances the prospects for sustainable development for all.

We must promote greater collaboration in trade, tourism and investment; infrastructure development; marine science and technology; sustainable fisheries; protection of marine environment; and, overall development of Ocean or Blue Economy.

To me the blue chakra or wheel in India's national flag represents the potential of Blue Revolution or the Ocean Economy. That is how central the Ocean Economy is to us.

For those who live by the ocean, climate change is not an issue of debate but a serious threat to existence. We must assume leadership in our region and call for a more concerted and fair global action to address the challenge of climate change.

Our Indian Ocean Rim Association can be an important instrument for pursuing our vision for a sustainable and prosperous future in the region.

We often define regional groupings around landmass. The time has come for a strong grouping around the Indian Ocean. We will pursue this with new vigour in the years ahead.

There can be no better place to host the Secretariat for IORA than Mauritius. I am pleased that the Secretary General is from India.

Five, those who live in this region have the primary responsibility for peace, stability and prosperity in the Indian Ocean.

But, we recognize that there are other nations around the world, with strong interests and stakes in the region.

India is deeply engaged with them. We do this through dialogue, visits, exercises, capacity building and economic partnership.

Our goal is to seek a climate of trust and transparency; respect for international maritime rules and norms by all countries; sensitivity to each other's interests; peaceful resolution of maritime issues; and increase in maritime cooperation.

We seek a future for Indian Ocean that lives up to the name of SAGAR – Security and Growth for All in the Region.

We should be inspired by Monsoon, which nurtures and connects us all in region.

We will strive to unite our region in partnership, as we were once in geography.

An Ocean that connects our world should become the pathway of peace and prosperity for all.

It is no coincidence that I speak of our hopes for the Indian Ocean Region in Mauritius.

Our partnership with Mauritius is among our strongest maritime relationships in this world.

Our partnership will grow. We will together build our capabilities. We will also train and patrol the seas together.

But, the foundation of this partnership is larger. It is our shared values and a common vision.

It is our willingness to look beyond our own domain to assume our responsibility for the region.

Mauritius is a key leader for a secure and sustainable future for the Indian Ocean.

We in India are proud to be your partner.

Thank you very much, and wish you all the best.

Appendix 2

DEFENCE OFFICERS ON DEPUTATION TO VARIOUS DIVISIONS OF MEA

(as on 30 Dec 2020)

BIMSTEC SAARC and Nalanda (BSN) Division

1.	Lt Col Ravinder Sharma (Indian Army)	OSD

BM [Bangladesh, Myanmar] Division

2.	Col VKS Tomar (Indian Army)	OSD

CPV [Consular Passport, Visa] Division

3.	Lt Col Ravi Raja (Indian Army)	OSD

DPA-II [Development Partner Administration] Division

4.	Lt Col Manish Mrinal (Indian Army)	OSD

DPA-III

5.	Col Mohit Nasa (Indian Army)	OSD

EG & IT [E-Governance & Information Technology] Division

6.	Lt Col Nishant Singh (Indian Army)	OSD
7.	Lt Col Shankar Singh (Indian Army)	OSD

Europe West Division

8.	Col Manjit Singh (Indian Army)	OSD [EW]

Gulf Division

9.	Cdr Madhvendra Singh (Indian Navy)	OSD [Gulf]

India-ASEAN Summit Secretariat

10.	Col US Rathore (Indian Army)	OSD [ASEAN ML-R]

11.	Col Digvijay Singh (Indian Army)	OSD [ASEAN ML-S]
Indian Ocean Region Division		
12.	Cdr Geo Mathews (Indian Navy)	OSD
PP & R [Policy Planning & Research] Division		
13.	Lt Col Vivek Malik (Indian Army)	Head [Boundary Cell]
PSP Division (Passport Seva Programme)		
14.	Lt Col AK Singh (Indian Army)	OSD
Southern Division		
15.	Cdr Arunava Bhattacharya (Indian Navy)	OSD
Sushma Swaraj Institute of Foreign Service		
16.	Lt Col Mayank Chaubey (Indian Army)	Consultant
UNP [United Nations Political] Division		
17.	Lt Col R Sharma (Indian Army)	OSD

Bibliography

Books

BK Sharma and Adil Rasheed (Eds.), *Indian Ocean Region: Emerging Strategic Cooperation, Competition and Conflict Scenarios* (New Delhi: Vij Books, 2015).

C Uday Bhaskar and Kamlesh K Agnihotri (Eds.), *Security Challenges Along the Indian Ocean Littoral-Indian and US Perspectives* (New Delhi: Matrix Publishers, 2011), p. 3.

Colin S Gray, *Air Power for Strategic Effect*, (Alabama, USA: Air University Press, 2014).

Côme Carpentier, *From India to Infinity* (New Delhi: Har-Anand Publications, 2012).

David Brewster, *India's Ocean: The Story of India's Bid for Regional Leadership* (London: Routledge, 2014).

Denis Venter, *India and Africa: Maritime Security and India's Strategic Interests in the Western Indian Ocean'*, in Iain Walker, Manuel João Ramos and Preben Kaarsholm (eds), *Fluid Networks and Hegemonic Powers in the Western Indian Ocean*, 2017, e-version at https://books.openedition.org/cei/469?lang=en.

Dhruv Katoch (Ed.), *India's Foreign Policy: Towards Resurgence* (New Delhi: Pentagon Press LLP, 2019).

Dhruv Katoch (Ed.), *Terrorism in Indian Ocean Region* (New Delhi: Pentagon Press, 2018).

Hemant Mahajan, *India's Coastal Security, Challenges, Concerns and Way Ahead* (Pune: Madhavi Publication, 2017).

John A Olsen (Ed.), *Air Power Reborn-the Strategic Concepts of John Warden and John Boyd* (Annapolis, Maryland: Naval Institute Press, 2015).

KM Panikkar, *India and the Indian Ocean- an essay on the influence of sea power on Indian history* (London: George Allen and Unwin Ltd., 1945).

Manoj Gupta, *The Indian Ocean Region: Maritime Regimes for Regional Cooperation* (New York: Springer, 2011).

PK Roy and Aspi Cawasji, *Strategic Vision 2030: Security and Development of ANC* (New Delhi: Vij Books, 2017).

R Sidda Goud and Manisha Mookherjee (Eds.), *China in Indian Ocean Region* (Hyderabad (India): Allied Publishers, 2015).

Robert Kaplan, *The Greater Indian Ocean in the 21st Century* (New York: Random House, 2011).

Sanat Kaul, *Andaman and Nicobar Islands-India's Untapped Strategic Assets* (New Delhi: Pentagon Press, 2015).

Sanjeev Sanyal, *The Ocean of Churn* (Gurugram: Penguin Books, 2016).

Shakti Sinha (Ed), *One Mountain Two Tigers*, (New Delhi: Pentagon Press LLP, 2020), p. 182.

Yogendra Kumar (Ed.), *Whither Indian Ocean Maritime Order* (Delhi: KW Publisher, 2017).

Periodicals/Journals/Magazines/Research Papers/Reports

AK Tiwary, "1971 Air War: Battle for Air Supremacy", http://www.indiandefencereview.com/spotlights/1971-air-war-battle-for-air-supremacy/.

Andrea Felber Seligman, "Ambassadors, Explorers, and Allies: A Study of African-European Diplomatic Relationships, 1400-1600", *CUREJ: College Undergraduate Research Electronic Journal*, University of Pennsylvania, http://repository.upenn.edu/curej/55, 23 March 2007.

Anil Chopra, "Aircraft Carrier Vs Shore Based Aircraft", *South Asia Defence Strategic Review*, Vol. 13, No. 4 (September-October 2019).

Arvind Dutta, "Role of India's Defence Cooperation Initiatives in Meeting the Foreign Policy Goals", *Journal of Defence Studies*, Vol. 3, No. 3 (July 2009).

Cristina Goulter, Harsh V Pant, "Realignment and Indian air power doctrine", *Journal of Indo-Pacific Affairs* (Fall 2018).

David Brennan, "China's Mix of Aggressive and Charming Coronavirus Diplomacy could Worsen Europe's Fractures, Report warns", https://www.ifri.org/en/espace-media/lifri-medias/chinas-mix-aggressive-and-charming-coronavirus-diplomacy-could-worsen

David Brewster, "An Indian Sphere of Influence in the Indian Ocean?", *Security Challenges*, Vol. 6, No. 3 (2010), pp. 1–20, JSTOR, www.jstor.org/stable/26459796.

David Brewster, "Covid contest in Indian Ocean region: India, China jostle for top spot", *The Interpreter*, https://www.lowyinstitute.org/the-interpreter/covid-contest-indian-ocean-region-india-china-jostle-top-spot

David Bulbeck, "Where River Meets Sea: A Parsimonious Model for Homo Sapiens Colonization of the Indian Ocean Rim and Sahul", *Current Anthropology*, Vol. 48, No. 2, University of Chicago Press, 2007, https://www.jstor.org/stable/10.1086/512988.

Douglas Harper, "Indian Ocean", Online Etymology Dictionary.

G Casale, "The ottoman 'discovery' of the indian ocean in the sixteenth century", *Seascapes: Maritime Histories, Littoral Cultures, and Transoceanic Exchanges*, University of Hawai'i Press, (December 01, 2007).

Geng Lixin, "Sino-Myanmar Relations: Analysis and Prospects", *The Culture Mandala*, Vol. 7, No. 2 (December 2006), http://www.international-relations.com/CM7-2WB/Sino-Myanmar.htm.

Harsh V Pant and Anant Singh Mann, "India's Malabar Dilemma", ORF Issue Brief issue No. 393 (August 2020), https://www.orfonline.org/wp-content/uploads/2020/08/ORF_IssueBrief_393_Malabar.pdf

IDSA, "Net Security Provider: India's Out-of-Area Contingency Operation", https://idsa.in/system/files/book/book_NetSecurityProvider.pdf.

"India Real GDP Growth, 2005–2020", *CEIC Data*, https://www.ceicdata.com/en/indicator/india/real-gdp-growth

Indian Defence Yearbook, "Modernisation Projects of the Indian Air Force", 2011.

K McPherson, "Cultural Exchange in the Indian Ocean Region", *Westerly Magazine*, Vol. 29, No. 4, (1984), https://westerlymag.com.au/cultural-exchange-in-the-indian-ocean-region/

Katariina Mustasilta, "From Bad to Worse?: The Impact(s) of Covid-19 on Conflict Dynamics", *European Union Institute for Security Studies (EUISS)*, 2020, www.jstor.org/stable/resrep25022.

Kishore Kumar Khera, "International Military Exercises: An Indian

Perspective", *Journal of Defence Studies*, Vol. 11, No. 3 (July-September 2017).

Lalit Kapur, "Geopolitical Lessons of COVID-19 for India", DPG policy papers, https://www.delhipolicygroup.org/publication/policy-papers/geopolitical-lessons-of-covid-19-for-india.html

N Boivin, Crowther A, M Prendergast, D Q Fuller, "Indian Ocean Food Globalisation and Africa", *African Archaeological Review*, Vol. 31, No. 4 (2014), https://doi.org/10.1007/s10437-014-9173-4

P Parthasarathi and G Riello, "The Indian Ocean in the Long Eighteenth Century", *Johns Hopkins University Press*, Vol. 48, No. 1 (Fall 2014).

Puneet Bhalla, "C-17: Enhancing India's Strategic Airlift capability", Scholar Warrior, CLAWS, Autumn 2013.

"Resolution 2532 (2020) Adopted by the Security Council on 1 July 20", https://undocs.org/en/S/RES/2532(2020)

"Ring of Fire", https://www.nationalgeographic.org/encyclopedia/ring-fire/

Sanjaya Baru, "India and the Indian Ocean Region: The New Geo-Economics", http://www.indiafoundation.in/india-and-the-indian-ocean-region-the-new-geo-economics-by-sanjaya-baru/

Shekhar Sinha, "Indian Navy: Opportunities in Changing Geopolitics of IOR", https://www.indiastrategic.in/2018/12/19/indian-navy-opportunities-in-changing-geopolitics-of-ior/.

Siddharth Singh and Deeksha Goyal, "Report: Impact of COVID-19 on the countries of Indian Ocean Region", https://indiafoundation.in/wp-content/uploads/2020/06/Impact-of-COVID-19-on-the-countries-of-Indian-Ocean-Region.pdf

Sumit Mukerji, "Dominating the High Ground Across the Indian Ocean Region", https://salute.co.in/dominating-the-high-ground-across-the-indian-ocean-region/,

Sushma Swaraj, "SAGAR – India's Vision for the Indian Ocean Region", https://indiafoundation.in/articles-and-papers/sagar-indias-vision-for-the-indian-ocean-region/.

Vikas Kalyani, "Australian view for Indian Ocean: significance for India", *Salute*, Vol. 11, No. 11 (July-August 2019).

Vikas Kalyani, "Global Affairs Expert Programme", *Salute*, Vol. 11, No. 2 (October-November 2018).

Vikas Kalyani, "Gulf Imbroglio: Changing Ge-Oil-itics", https://

indiafoundation.in/articles-and-papers/gulf-imbroglio-changing-ge-oil-itics/.

Warfare History Network, "FACT: The Last Aircraft Carrier vs Aircraft Carrier Battle Was Decades Ago: A World War II showdown", 17 November 2019, https://nationalinterest.org/blog/buzz/fact-last-aircraft-carrier-vs-aircraft-carrier-battle-was-decades-ago-96846.

World Economic Forum report, "COVID-19 hit commodities hard - these 8 charts show what we can expect", https://www.weforum.org/agenda/2020/11/commodity-markets-economics-charts-oil-energy-metals-gold-food-covid-19-coronavirus-pandemic/

Articles/Newspaper/News on Internet

Ajai Shukla, "Navy invites neighbours to join Indian Ocean monitoring facility in Gurgaon", *Business Standard*, 23 December 2018, https://www.business-standard.com/article/defence/navy-invites-neighbours-to-join-indian-ocean-monitoring-facility-in-gurgaon-118122300007_1.html.

Amin Bagheri, "Is Coronavirus a Bioweapon?", 03 July 2020, https://www.newsgram.com/coronavirus-bioweapon/

Angad Singh, "Focus Beyond Fighters", https://www.orfonline.org/expert-speak/focus-beyond-fighters-56374/.

ANI, "Defence Space agency to come up at Bengaluru", 15 May 2019, https://economictimes.indiatimes.com/news/defence/defence-space-agency-to-come-up-at-bengaluru/articleshow/69345697.cms?from=mdr.

ANI, "India has extended assistance to over 150 countries in fight against COVID-19: PM Modi", https://health.economictimes.indiatimes.com/news/policy/india-has-extended-assistance-to-over-150-countries-in-fight-against-covid-19-pm-modi/77028422

Ankit Panda, "Australia Returns to the Malabar Exercise", https://thediplomat.com/2020/10/australia-returns-to-the-malabar-exercise/

Ankit Panda, "South China Sea: What China's First Strategic Bomber Landing on Woody Island Means", *The Diplomat*, 22 May 2018.

Bertil Lintner, "China eyes a Covid-19 edge in the Indian Ocean", *Asia Times*, https://asiatimes.com/2020/04/china-eyes-a-covid-19-edge-in-the-indian-ocean/

Bloomberg News, "Chinese aircraft breach Taiwan airspace; raise risk of clash in the region", https://www.business-standard.com/article/international/chinese-aircrafts-breach-taiwan-airspace-raise-risk-of-clash-in-the-region-120092100320_1.html

Brilliantwiki2, https://commons.wikimedia.org/wiki/File:IRNSS_Coverage_Polygon.png#file.

BS Dhanoa, "Air Ambulance landmark achievement in trauma surgery", 25 June 2017, https://www.financialexpress.com/india-news/air-ambulance-landmark-achievement-in-trauma-surgery-indian-air-force-chief/735351/.

"C-130 J-Specification and Performance", 16 December 2019, http://www.globalsecurity.org/military/systems/aircraft/c-130j-specs.htm

"China's Long March 2C Rocket Launches Military Surveillance Satellites Into Orbit", https://spacewatch.global/2020/03/chinas-long-march-2c-rocket-launches-military-surveillance-satellites-into-orbit/

Constantino Xavier, "Getting Back Home, Safely", 08 February 2017, https://carnegieindia.org/2017/02/08/getting-back-home-safely-pub-67974.

"Corona Tracker: India Overview", https://www.coronatracker.com/country/india/ accessed on 30 October 2020.

"Covid-19 Coronavirus Pandemic", https://www.worldometers.info/coronavirus/ accessed on 30 October 2020.

Debanish Achom, https://www.ndtv.com/india-news/as-biggest-vaccine-maker-india-will-help-world-overcome-covid-pm-narendra-modi-to-un-general-assembly-2301496

Dinakar Peri, "Weaponisation by China a threat: U.S. Admiral", *The Hindu*, 30 October 2019.

Explained desk, "Explained: how has covid-19 affected the global economy", https://indianexpress.com/article/explained/explained-how-has-covid-19-affected-the-global-economy-6410494/

"Explained: South China Sea dispute", 16 February 2019, https://www.scmp.com/week-asia/article/2186449/explained-south-china-sea-dispute

Farzin Nadimi, "A China-Pakistan Base Deal Could Put Iran on the Back Foot", 27 March 2018, https://www.washingtoninstitute.org/policy-analysis/view/a-china-pakistan-base-deal-could-put-iran-on-the-back-foot.

Geeta Mohan, "China's growing territorial aggression amid corona virus crisis", https://www.indiatoday.in/mail-today/story/china-growing-territorial-aggression-amid-covid-crisis-1692787-2020-06-26

Harsh V Pant, "India's challenge in the Indian ocean region", https://www.orfonline.org/expert-speak/indias-challenge-in-the-indian-ocean-region.

Hasan Alhasan, https://www.asiatimes.com/2019/09/opinion/big-asian-players-too-quiet-on-threats-to-gulf-security.

How "How CDS can integrate the services", *The Sunday Tribune*, 29 December 2019.

HS Ahluwalia, "IAF in Maritime Operations", http://bharat-rakshak.com/cms/srr/64-17.htm.

Huma Siddiqui, "Coronavirus Pandemic: India lends helping hand to it's neighbour Myanmar; send medical supplies", https://www.financialexpress.com/defence/coronavirus-pandemic-india-lends-helping-hand-to-its-neighbour-myanmar-send-medical-supplies/1948830/ (Accessed 30 September 2020)

"India calls for UN force to deter pirates off Somalia", *Times of India*, 13 November 2008, https://timesofindia.indiatimes.com/india/India-calls-for-UN-force-to-deter-pirates-off-Somalia/articleshow/3710102.cms.

"Indian Air Force Contingent Leaves for Sudan", 18 January 2010, https://www.airforce-technology.com/news/news74514-html/.

"Indian Ocean", www.etymonline.com/word/indian%20ocean.

Kabir Taneja, "Reading a potential Iran–China strategic agreement from India", https://www.orfonline.org/expert-speak/reading-a-potential-iran-china-strategic-agreement-from-india/

Karim Raslan, "OPINION: In the South China Sea, ASEAN needs to stand united", 14 January 2020, https://news.abs-cbn.com/blogs/opinions/01/14/20/opinion-in-the-south-china-sea-asean-needs-to-stand-united.

Katharina Buchholz, "Coronavirus Slump Continues For GDP Growth", https://www.statista.com/chart/18095/quarterly-gdp-growth-predicted-growth-selected-industrialized-nations-oecd/

Koduvayur Varsha, David Daoud, "Welcome to a Brand-New Middle East", https://foreignpolicy.com/2020/09/30/israel-uae-bahrain-palestinians-peace/

Li Yuan, "Coronavirus Crisis Shows China's Governance Failure", https://www.nytimes.com/2020/02/04/business/china-coronavirus-government.html?auth=login-google

Lindsey Ford, Julian Gewirtz, "China's post-COVID aggression is reshaping Asia", https://www.thehindu.com/opinion/lead/chinas-post-covid-aggression-is-reshaping-asia/article32095103.ece

Lora Jones, Daniele Palumbo and David Brown, "Coronavirus: A visual guide to the economic impact", *BBC News*, https://www.bbc.com/news/business-51706225#:~:text=Risk%20of%20recession&text=But%20the% 20IMF %20s ays%20that Great% 20Depression%20of%20the%201930s.&text=Recovery%20in%20big%2C%20services%2Dreliant, to%20be%20a%20slow %20process.

Manjeet Negi, https://www.indiatoday.in/india/story/navy-tracks-down-chinese-nuclear-submarines-operating-close-to-indian-waters-1599555-2019-09-16.

Mark McDonald, "China Sends Naval Task Force on Anti-Piracy Mission", *The New York Times*, 26 December 2008.

Martand Jha, "Why the Indian Ocean Conference is Becoming Important", https://www.telegraphindia.com/opinion/why-the-indian-ocean-conference-is-becoming-important/cid/1670382.

Niranjan Marjani, "India's Indian Ocean Diplomacy in the COVID-19 Crisis", *The Diplomat*, https://thediplomat.com/2020/04/indias-indian-ocean-diplomacy-in-the-covid-19-crisis/

NK Bhatia, "Iran-China strategic agreement and its implications for India", https://www.financialexpress.com/defence/iran-china-strategic-agreement-and-its-implications-for-india/2022453/

NM Dhavle, "Chetak Dettachment to Batticoloa", 22 April 2013, http://www.bharat-rakshak.com/IAF/History/1970s/1113-Dhavle.html.

"Overseas Operations of the Indian Air Force", 10 July 2013, http://www.bharatdefencekavach.com/news/indian-air-force/overseas-operations-of-the-indian-air-force/37711.html.

Pallav Agarwal, "70 years of peacekeeping: Achievements, challenges and need for reforms", 10 December 2018, https://www.orfonline.org/expert-speak/70-years-of-peacekeeping-achievements-challenges-and-need-for-reforms-46135/.

Plamen Tonchev, "The Belt and Road After COVID-19", https://thediplomat.com/2020/04/the-belt-and-road-after-covid-19/ accessed on 11 June 2020.

Pooja Bhatt, "Achieving Indian Air Dominance in the Indian Ocean Region", 17 May 2018, https://southasianvoices.org/achieving-indian-air-dominance-in-the-indian-ocean-region.

"Preserve organic unity of Indian Ocean Region, advance cooperation – Swaraj", *The Island*, 03 September 2017, http://www.island.lk/index.php?page_cat=article-details&page=article-details&code_title=171146.

Priyanka Deo, "Steps India must take for strong foreign relations post COVID-19", https://www.wionews.com/opinions-blogs/steps-india-must-take-for-strong-foreign-relations-post-covid-19-304818

PTI, "India's fight against Covid-19 driven by its people: PM Modi", https://theprint.in/india/indias-fight-against-covid-19-driven-by-its-people-pm-modi/450183/

PTI, "Indian Army further strengthens dominance in strategic heights in eastern Ladakh", https://www.deccanherald.com/national/indian-army-further-strengthens-dominance-in-strategic-heights-in-eastern-ladakh-885596.html

Raghbendra Jha, "India's Response to the Coronavirus Crisis", *Australian Outlook, AIIA*, http://www.internationalaffairs.org.au/australianoutlook/indias-response-to-the-coronavirus-crisis/

Rajeswari Pillai, "What's in the Growing Russia-China-Iran Trilateral Convergence?", *The Diplomat*.

Rajeswari Pillai and Nagendra Narayan Prasad, "Creation of a defence space agency: A new chapter in exploring India's space security", https://www.orfonline.org/expert-speak/creation-of-a-defence-space-agency-a-new-chapter-in-exploring-indias-space-security/.

Rekha Dixit, "India provides $250 million aid to Maldives to mitigate COVID-19 impact", https://www.theweek.in/news/india/2020/09/20/india-provides-250-million-aid-to-maldives-to-mitigate-covid-19-impact.html

"Relief ops 'Castor' and 'Rainbow' launched", *Times of India*, 27 December 2004, https://timesofindia.indiatimes.com/india/Relief-ops-Castor-and-Rainbow-launched/articleshow/972740.cms.

Rishab Gulati, "The truth behind China's Ladakh Spy sats and how it planned Galwan heist", https://www.newsx.com/national/the-truth-behind-chinas-ladakh-spy-sats-and-how-it-planned-galwan-heist.html

Rogoway Tyler, https://www.thedrive.com/the-war-zone/30020/china-deploying-a-weaponless-stealth-drone-on-its-carriers-makes-perfect-sense.

Sameer Pradhan, "Fighting COVID-19: IAF'S response to the Coronavirus threat", https://www.financialexpress.com/defence/fighting-covid-19-iafs-response-to-the-coronavirus-threat/1991320/

Shekhar Sinha, "The Indian Navy and the Geopolitics of the Indian Ocean", https://sldinfo.com/2018/12/the-indian-navy-and-the-geopolitics-of-the-indian-ocean/.

Shishir Gupta, "India draws up Rs 1 billion Covid-19 medical assistance plan, targets 90 countries", https://www.hindustantimes.com/india-news/india-amps-up-covid-19-medical-assistance-plan-targets-to-reach-90-countries/story-0X1H8z1Zqi9piw6n4FDu8J.html

Shrey Kapoor, "Why was the Indian Ocean named after India?", https://www.techphlie.com/2016/05/why-was-indian-ocean-named-after-india.html.

Sivapriyan, "Sukhoi-30 MKI jets inducted in Thanjavur air base", https://www.deccanherald.com/national/south/sukhoi-30-mki-jets-inducted-in-thanjavur-air-base-796480.html.

Srinivasa Rao Apparasu, *Hindustan Times*, https://www.hindustantimes.com/india-news/not-at-war-with-china-but-fully-prepared-says-air-force-chief-bhadauria/story-Lhs3mpn1DfBIGcWkmMTUnM.html

Sudhanshu Tripathi, https://timesofindia.indiatimes.com/blogs/abhiram/is-corona-virus-a-new-bio-weapon/

Sutirtho Patranobis, https://www.hindustantimes.com/world-news/china-launches-its-first-amphibious-warship/story-j5V2P6feCuHFGJYIgHlRTO.html.

The Economic Times, "PM Modi hails South Sudan evacuation operation", 17 July 2016, https://economictimes.indiatimes.com/news/politics-and-nation/pm-modi-hails-south-sudan-evacuation-operation/articleshow/53225885.cms?utm_source=contentofinterest&utm_medium=text&utm_campaign=cppst.

The Economist, "Would-be autocrats are using covid-19 as an excuse to grab more power", https://www.economist.com/international/2020/04/23/would-be-autocrats-are-using-covid-19-as-an-excuse-to-grab-more-power

"The Indian Air Force in Congo", http://www.bharat-rakshak.com/IAF/History.html.

"The Sri Lankan Interlude 1987-90", http://www.bharat-rakshak.com/IAF/History/1987IPKF.html.

"Updates: How Navy Commander Abhilash Tomy was rescued", 24 September 2018, https://www.indiatoday.in/india/story/live-updates-abhilash-tomy-stranded-indian-navy-officer-rescued-safely-1347408-2018-09-24.

Valentina Romei, John Reed, "The Asian century is set to begin", https://www.ft.com/content/520cb6f6-2958-11e9-a5ab-ff8ef2b976c7

Vikas Kalyani, "The Indian Air Force and an increasing HADR mandate", https://www.orfonline.org/expert-speak/indian-air-force-increasing-hadr-mandate-65992/

"WHO declares Covid-19 outbreak a pandemic", https://www.pharmaceutical-technology.com/news/who-declares-covid-19-pandemic/, 12 March 2020.

"World Air Forces 2020", Flight Global in association with Ruag, https://www.flightglobal.com/reports/world-air-forces-2020/135665.

Zachary Williams, "China's Tightening Grasp in the South China Sea: A First-Hand Look", *The Diplomat*, https://thediplomat.com/2020/06/chinas-tightening-grasp-in-the-south-china-sea-a-first-hand-look/

Speeches and Interview

BS Dhanoa in a lecture at Vivekananda International Foundation, April 2018.

Narendra Modi, Indian PM, https://www.pmindia.gov.in/en/news_updates/text-of-the-pms-remarks-on-the-commissioning-of-coast-ship-barracuda/ 12 March 2015.

Personal interview with Mr Amit Narang, JS IOR, MEA on 16 December 2019.

RKS Bhadauria, Chief of Air Staff, IAF, Raisina Dialogue 2020, 15 January 2020.

Sushma Swaraj, EAM, GoI at Indian Ocean Conference 2017 (http://indiafoundation.in/sagar-indias-vision-for-the-indian-ocean-region dated 29/11/18).

Sushma Swaraj, https://www.mea.gov.in/Speeches-Statements.htm?dtl/ 30327/ Remarks+by+External+ Affairs+Minister+at+the+3rd+Indian+Ocean+Conference+Vietnam+August+27+2018.

Government Websites/Documents

"Aid to Civil Power", https://indianairforce.nic.in/content/aid-civil-power.

Annual Reports (2011–12 to 2015–16), Ministry of Defence.

Basic Doctrine of the IAF, 2012.

https://mea.gov.in/circulars-notifications.htm.

https://mea.gov.in/divisions.

IAF Contingent in Somalia, https://indianairforce.nic.in/content/somalia.

"IAF in UN Missions", https://indianairforce.nic.in/content/indian-aviation-contingent-i-iac-i.

Indian Regional Navigation Satellite System (IRNSS) : NavIC, https://www.isro.gov.in/irnss-programme

LIMA 2019, https://indianairforce.nic.in/content/lima-2019.

"Operation Cactus", https://indianairforce.nic.in/op-cactus.

"Operation Insaniyat", Press Release on 14 September 2017 https://mea.gov.in/pressreleases.htm?dtl/28944/ Operation_Insaniyat__Humanitarian_assistance_to_Bangladesh_on_account_of_influx_of_refugees.

Population of Overseas Indians (Compiled in December, 2018), https://mea.gov.in/images/attach/NRIs-and-PIOs_1.pdf.

Press Information Bureau, Government of India, "INS 'Baaz' – Commissioned by Admiral Nirmal Verma as First Naval Air Station in Nicobar Group of Islands", 31 July 2012.

Press Trust of India, "India and other countries must fight terrorists in Afghanistan at some point, says Donald Trump", *First Post*, 22 August 2019.

Remya Lakshmanan and Mishika Nayyar, "Personal Protective Equipment in India: An INR 7,000 Cr industry in the making", https://www.investindia.gov.in/siru/personal-protective-equipment-india-INR-7000-cr-industry-in-the-making

"Towards Self Reliance in Navigation-IRNSS", https://www.isro.gov.in/towards-self-reliance-navigation-irnss.

"Vice Chief of the Air Staff Commissions Pilot Project on Modernisation of Airfield Infrastructure (MAFI)", 26 March 2014, https://pib.gov.in/newsite/PrintRelease.aspx?relid= 104691.

Year End Review - 2017, 2018: Ministry of Defence, Press Information Bureau.

Index

*